SEATTLE GEOGRAPHIES

SEATTLE GEOGRAPHIES

EDITED BY

MICHAEL BROWN
RICHARD MORRILL

A Samuel and Althea Stroum Book

UNIVERSITY OF WASHINGTON PRESS
Seattle & London
in association with the
DEPARTMENT OF GEOGRAPHY
University of Washington

Seattle Geographies is published with the assistance
of a grant from the Samuel and Althea Stroum
Endowed Book Fund.

The book also received generous support from the
University of Washington's College of Arts and Sciences,
Department of Geography, Office of the Provost, and
Office of Undergraduate Affairs, as well as a number
of private donors.

UNIVERSITY OF WASHINGTON PRESS

PO Box 50096, Seattle, WA 98145, USA

www.washington.edu/uwpress

DEPARTMENT OF GEOGRAPHY

University of Washington, Seattle, WA 98195

http://depts.washington.edu/geog

Printed and bound in the United States of America

Designed by Ashley Saleeba

Composed in Meta Serif and Interstate

FRONTISPIECE: Elliott Bay and Seattle skyline as seen from Alki.
Photograph © 2009 by Andrew Gorhoff.

LIBRARY OF CONGRESS CATALOGING-IN-PUBLICATION DATA

Seattle geographies / edited by Michael Brown and Richard Morrill.

   p. cm. — (A Samuel and Althea Stroum book)

Includes bibliographical references and index.

ISBN 978-0-295-99091-0 (pbk. : alk. paper)

1. Human geography—Washington (State)—Seattle.

2. Seattle (Wash.)—Geography.

I. Brown, Michael, 1966– II. Morrill, Richard L., 1934–

GF504.2.S43 2011

304.209797'772—dc22   2010050572

The paper used in this publication meets the minimum
requirements of American National Standard for Information
Sciences—Permanence of Paper for Printed Library Materials,
ANSI Z39.48-1984.∞

This book is dedicated to the place we call home.

CONTENTS

ACKNOWLEDGMENTS

The idea for this book emerged at one of our weekly meetings over beer at the Roanoke Tavern on North Capitol Hill in Seattle. Despite our generation gap, different methodological approaches, and sometimes very different research interests, we have been meeting every week for years to discuss politics, Seattle, and any other subject that might catch our attention. Our chats have lead to several research projects and papers, but this particular project is our favorite. We would like to begin by thanking the folks at the Roanoke Tavern—especially Dennis—for all the hospitality and fine beer.

We also thank our colleagues and students in the Geography Department at the University of Washington, who generously contributed to the book. Working toward community in this current, financially strapped academic climate can be tough, when pressures are mounting to take on only work that is necessary and efficient. We are privileged to have wonderful colleagues and students who agree that this kind of community-building and geographic writing with a common voice are worthwhile. We would especially like to thank several colleagues who were helpful and supportive at key stages of the process, including Sue Bernhardt, William Beyers, Mark Ellis, J. W. Harrington, Judy Howard, and Suzanne Withers. We thank especially department chair Katharyne Mitchell for generously paying for the reproduction of prints and for securing permissions. There is precious little money for projects like this one, and we could not have included many of the photographs without this support.

Financial support for the color printing of this book was provided by the University of Washington's Office of the Provost, Undergraduate Education Unit, College of Arts and Sciences, and Geography Department, as well as by donations from William B. Beyers, Victoria Lawson, Richard Morrill, and Michael Brown. We thank the owners of the images for granting us permission to reproduce them here. We have made every effort to obtain permission from owners of every figure in the book.

Sean Wang was an indefatigable research assistant, and we thank him for all his hard work. Mike Babb's cartographic skill was invaluable, as was Josef Eckert's help. Andrew Childs also helped us proofread and edit the final manuscript. At the University of Washington Press, Beth Fuget was a supportive and insightful editor and Director Pat Soden was an early and enthusiastic supporter of the project.

Finally, we would like to thank Joanne Morrill and Jonathan Cronin for their support, love, and encouragement.

Michael Brown and Richard Morrill
The Roanoke Tavern, Seattle
September 2010

SEATTLE GEOGRAPHIES

ONE

INTRODUCING SEATTLE GEOGRAPHIES

Richard Morrill, William Beyers, and Michael Brown

Seattle is at once an extremely easy and rather difficult place to understand. It shares many characteristics with other metropolitan areas but is unique and full of contradictions. Its economy is postindustrial and yet manufacturing still is strong. Its politics are liberal but social controls abound. Its culture is cosmopolitan, with elements of provincialism. Geography—as an academic discipline—is similarly paradoxical, sharing topics with all the other social and physical sciences and yet unique in its synthetic approach to the world. Geographers typically cross disciplinary boundaries, leaving many confused about just what it is that unites the field. Because of these complexities of subject and discipline we want to introduce Seattle and its region in terms of both human and physical geographies, with reference to the nature of geography as a science.

Seattle as a "Place"

What is distinctive about the Seattle region? The city is big but not a giant, big enough to have attained recognition as a world player but small enough for its environment-loving citizens to reach the wilderness within an hour. It is a major port in the beautiful enclosed Puget Sound, and it is dependent upon trade. It is defined by its iconic global firms—historically, by Boeing aircraft, PACCAR trucks, Weyerhaeuser forest products (the resource base is not quite gone), and, in the twenty-first century, by Microsoft (and allied and competing "digital" firms), Amazon, Costco, Nordstrom, REI in retail, and Starbucks. There also was

(opposite)

1.1 Downtown Seattle skyline.

1.2 The Space Needle at night.

a mega-bank, Washington Mutual, and there still is the world-class University of Washington and related biomedical offspring, such as the Fred Hutchinson Cancer Care Center. As global cities go, Seattle probably just makes the ranks of the popular conception of a "world city" (see chapter 3). It is a little brash and youthful, barely 150 years old, and broke into the big leagues only in the last thirty years or so. The metropolis has experienced (if not always enjoyed) rapid growth, but that growth has been volatile, punctuated by loss and recession.

Seattle's demographics are unusual (see chapter 6). Today, migrants make up half or more of its population. Its Asian American population share is high and its African American share low compared to other American cities. It has a relatively large share of nonfamily households. The region's culture is exceptionally progressive, with a high degree of social activism and tolerance among the citizens. Religious adherence is comparatively low. Seattle epito-

mizes the modern, tolerant, secular city. Still, it has rarely been contented. Portland, Oregon, and Vancouver, B.C., claim to be "more cosmopolitan," so our leaders' aspiration has always been to outdo these rivals, to mobilize us to greater heights—literally, as with the Space Needle (fig. 1.2) and the Columbia Tower, and figuratively, with world-class distinction.

Seattle has a dual role to play in the new millennium. It is one of the twenty or so regional capitals of the United States. At 12 million population, the Greater Seattle region is of intermediate size, comparable to the Minneapolis region. As a metropolis of 3.5 million, it is also of intermediate importance. Seattle is at a third tier in the national hierarchy—subordinate to the major world cities of New York and Los Angeles and also to the historic "capital of the

West," San Francisco, which has a population of more than 7 million and is still the home of the Federal Reserve district. The Pacific Northwest (see fig. 1.9), for which Seattle is the capital, consists of the states of Washington, Oregon, Alaska, Idaho, and western Montana. Seattle exercises trade, finance, and service dominance in this region, exemplified most visibly by Seattle-Tacoma International Airport, the health and medical complexes, the Port of Seattle, professional sports, and the arts. That is, Seattle is the hub or control center of a dependent hinterland.

Seattle has the highest per capita income and wealth in the Pacific Northwest. There are two reasons for this: (1) the large size of our labor force and internal market brings "agglomeration and scale" benefits—the widest variety of activities in one place plus lower unit costs with higher volumes—giving us more self-sufficiency and efficiency than smaller places; (2) a favorable balance of trade with the hinterland means that the rest of the region has to come to Seattle for the highest level of goods and services that they cannot provide for themselves. It is possible that as much as one-fifth of the wealth in Seattle is simply a result of people coming to the city and spending and investing at a rate higher than people from Seattle spend and invest in the hinterland. Note that this imbalance is mostly on the private side: Seattle area taxpayers subsidize the hinterland on some public expenditures, such as highways and schools.

Seattle's other major role is as a specialized global producer and trader. Not only is a large part of our production, probably one-third, exported to other regions of the country and to other nations, but we also serve (for a fee) as the export and import gateway for large amounts of goods flowing to and from the rest of the country and East Asia and beyond. Seattle is the American city most dependent on

1.3 Ships on Puget Sound with Mount Rainier in the background.

foreign exports. Foreign trade comprises about 25 percent of the regional economy, which poses some risk of volatility. Given that the United States has an enormous deficit imbalance of trade with the rest of the world, however, Seattle provides a big service to the American economy by having such a large positive balance of trade.

The Seattle region has many definitions (compare chapters 3 and 4, for example). At the minimum, it is the city limits of Seattle, but if Alaska and western Montana depend on Seattle and Seattle has dams and power facilities in Idaho and landfills in Oregon, then the "Seattle region" is the entire Pacific Northwest. So, in this book we are arbitrarily defining the Seattle region as primarily the four central Puget Sound counties: King, Pierce, Snohomish and Kitsap. At times, we may emphasize the inner urban core, and at other times not hesitate to extend beyond the four counties, when Seattle's reach does just that. Geographers do research at different spatial scales, from the neighborhood to the city to the region. We sometimes differ in our definitions of a region (see chapters 2 and 4), and some of us study relations between places (see chapter 3, which explores the city's situation globally). The Seattle region is what its people know as a zone of everyday interaction of people in space.

If "geography is destiny," as some claim, the position of a place relative to the wider scale of the nation and the world and to the local resources and opportunities would seem critical to its future. With respect to the continental United States and its twin seats of power, New York City and Washington, D.C., Seattle is hopelessly remote. Modern transportation and communication have reduced this marginality, but have not eliminated all the costs of movement. Seattle is even peripheral with respect to the loci of power in the West, Los Angeles and San Francisco, on which we are far more dependent than some people may realize.

At the global scale, however, Seattle's situation proved favorable to its rapid growth at the end of the nineteenth century. As the port closest to China, it prospered with the silk trade (see chapter 2). Seattle, Tacoma, and Everett also had ready rail access south, for shipments of lumber for San Francisco's growth and rebuilding after fires and earthquakes, as our forest resources were unsurpassed in quality and access at the time. Most importantly, the 1897 discovery of gold in the Yukon and Seattle's proximity to Alaska triggered the city's growth, enrichment, and ultimate domination of the Pacific Northwest. Symbolized by Alaska Airlines' headquarters in Seattle and by the funneling of much trade through Seattle, Alaska has never escaped Seattle's strategic intervening position. Puget Sound may have been remote from eastern centers of power, but its extreme northwestern location necessarily made it a vital military and defense outpost, which resulted in an enduring strong military presence and contributed to Seattle surpassing rival Portland in importance.

In the Pacific Northwest, Seattle's situation is markedly inferior to that of Portland's, with the latter having far easier access to the interior of the region. Even attracting the transcontinental railroads did not overcome this relative weakness for Seattle. Instead, its current dominance is a story of overcoming geographic constraints by entrepreneurship and politics.

At the smaller scale of Puget Sound, Seattle's position is indeed central but in no way made the city significantly more accessible to the subregion than its early rivals of Tacoma or Everett. Seattle's dominance, despite the transcontinental railroads reaching Tacoma and Everett first, is a fascinating story of human and political struggles.

Over time Seattle's remote but permissive environment afforded opportunities for development and trade and transformed it into a surprisingly powerful world player, considering its modest size by world standards. This role was achieved through a history of entrepreneurship, of often enlightened economic and political leadership, and above all by a population—and its institutions and culture—that encouraged and enabled cooperation as well as competition. The battle, so to speak, between entrepreneurship, boosterism, and dependence on large corporations and the countervailing forces for addressing inequality and social concerns exemplifies the ongoing story of Seattle.

THE PHYSICAL GEOGRAPHY

William Beyers

Seattle is located in a stunning physical setting, which is, in turn, part of the very high quality physical environment that constitutes Washington State. Here we look at aspects of the landforms, climate, and vegetation of an extremely diverse physical environment, possibly the most diverse of any state in the conterminous United States.

Landforms

Washington State is divided by the Cascade Mountains into the cool, maritime climate–dominated landscape of Western Washington and the more continental climate–dominated landscape of Eastern Washington. West of the Cascade foothills, Western Washington is composed of the Olympic Mountains, the Coast Range, and the Puget-Willamette lowlands.

The Olympic Mountains rise to the west of Seattle, on the Olympic Peninsula. Their elevation reaches 7,954 feet on Mount Olympus, with many other summits in this range over 7,000 feet. These mountains rise sharply from sea level on their east and north sides, from Hood Canal and the Strait of Juan de Fuca. Valleys deeply incise this range, with major rivers, such as the Elwha, Hoh, Quinault, Dosewallips, Dungeness, and Bogachiel, penetrating at low levels into the heart of the range.

Southwest Washington contains part of the Coast Range (which extends through western Oregon), a hilly region, with the bays of Grays Harbor and Willapa Harbor on the ocean.

1.4 Looking west toward the Olympic Mountain Range.

The Puget Lowland runs from the Canadian border to the Columbia River. The northern part of this region contains Puget Sound (part of the Salish Sea) and the San Juan Islands, while the southern part has valleys and low hills between the Cascade Mountains and the Coast Range running from Olympia to Vancouver. The northern part of the Puget Lowland was covered by Pleistocene glaciation, which left the morainal hills and valleys that constitute the majority of the area's topography.

The Cascade Mountains run from Canada to the Columbia River in Washington State, and the Columbia is the only river that penetrates through this range in Washington and Oregon. The range contains volcanic peaks, such as Mount Rainier, Mount Baker, and Glacier Peak (all of which are visible from Seattle), as well as hundreds of rocky peaks with extensive glaciation. The section of the Cascade Mountains from Stevens Pass to the Canadian border is the most rugged mountain range in the coterminous United States, with local relief of more than a vertical mile in many places. The region due east of Seattle, known as the Alpine Lakes, is slightly less rugged but is dotted with hundreds of cirque basin lakes and larger lakes created by glacial activity. The southern Washington Cascades are hilly, with fewer high peaks and glaciers than found in the North Cascades.

East of the Cascade Mountains the landscape can be divided into the Columbia Basin and an extension of the Northern Rocky Mountains. The Columbia Basin extends from the east slopes of the Cascade Mountains to the west slopes of the Rocky Mountains (about at the Washington-Idaho border). It is bordered on the south by the Columbia River, and in extreme southeast Washington by the Blue Mountains. The mighty Columbia River and its tributar-

1.5 View of Interstate 90 and the Cascade Mountain Range.

ies are the primary waterways in this interior region—the Columbia extends into Canada and east into Idaho and Montana, and the Snake River tributary extends through southern Idaho.

The Columbia Basin landscape has been heavily influenced by two factors: enormous basalt flows and catastrophic floods during the last Ice Age. Tabular basalt is found across the Columbia Basin province and extends into eastern Oregon and southern Idaho. These basalt flows have been warped into anticlines and synclines in the southwest part of the region (the "Yakima Folds"), were scoured by glaciation in the northwest (the Waterville Plateau), and were converted into dramatic watercourses in the channeled scablands and the great coulees (such as Grand Coulee and Dry Falls). The Ice Age floods, coming from Lake Missoula (east of Spokane into Montana), caused riverlike features on an enormous scale, from Spokane west to the current location of the Columbia River, including down to its confluence with the Pacific Ocean. During the Ice Age, winds blew soil from the central part of the Columbia Basin eastward into what is today the Palouse (in southeast Washington), a region of dunelike hills composed of loess soils that are hundreds of feet thick. Today this area is prime wheat land.

To the north of the Columbia Basin, in northeast Washington, the landscape is more mountainous, with hills and valleys aligned in a north-south manner, dominated by the Columbia, Okanogan, Methow, and Pend Orielle rivers.

Climate

With its rich topographic character, Washington also has extremely diverse climates and many microclimates. The area west of the Cascades is strongly influenced by maritime forces off the Pacific Ocean. Across the state, the Pacific Ocean significantly influences the weather and the climate, but in Eastern Washington northerly weather systems from Canada and the Arctic have a greater impact.

The state as a whole has a winter maximum in precipitation. The level of precipitation varies enormously—from temperate rain forest levels in the range of 120–140 inches annually on the west slopes of the Olympics to as low as 6 inches annually in parts of the rain shadow east of the Cascades. Seattle gets about 37 inches of rainfall annually, with measurable precipitation on about two hundred days per year. Along the Pacific Coast, annual rainfall levels are in the 70-to-90-inch range. In the Puget-Willamette lowland, annual rainfall levels are in the 35–50 inch range, while on the west slopes of the Cascades, precipitation levels rise to 75–100 inches. However, there is a significant rain shadow to the northeast of the Olympic Mountains, in the region near Sequim, Port Townsend, the San Juan Islands, northern Whidbey Island, and western Skagit County. To the east of the Cascades, the rain shadow is prevalent, and precipitation falls quickly, to 20–40 inches approximately fifty miles east of the Cascade crest, and then to only 5–15 inches near the Columbia River. Toward the Rocky Mountains, precipitation levels increase again, to about 15 inches in the Palouse, which is ideal for wheat culture.

Much of the precipitation in higher elevations falls in the form of snow in the winter. Snowpacks peak in water content in May or June; accumulations at higher elevations usually start in October and typically do not melt off until July or August.

Temperatures are also highly variable across the state in all seasons. Western Washington lowlands have mild temperature conditions, influenced by maritime proximity. Winter temperatures are usually above freezing in the lowlands; while in the summer, daytime highs are usually in the 70s. The mountains have much greater temperature variation, with generally freezing temperatures above 3,000 feet in winter, and daytime highs in the 70s in summer months. Eastern Washington has much more temperature variation than Western Washington. Winter low temperatures are generally below freezing, while summer high temperatures often reach 100 degrees F.

There are also major differences in cloud cover and frequency of sunshine across Washington State. Eastern Washington and locations in Western Washington in the rain shadow of the Olympics have a much higher share of

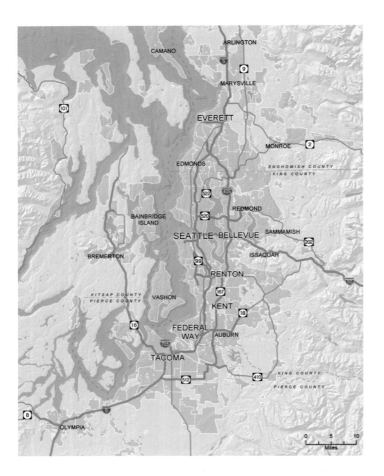

1.6   Seattle in the regional metropolitan context.

1.7   The city of Seattle and its neighborhoods.

the days with sunshine than in the mountains or elsewhere west of the Cascades. Cloud cover from storms off the Pacific and fog are common in Western Washington.

Vegetation

The natural vegetation of Washington State is as diversified as the climate and physical environment. The area west of the Cascades has naturally lush, temperate, coniferous forests, while the area east of the Cascades was naturally dominated by desertlike vegetation. Rainfall levels and climate are strongly related to vegetative differences across the state. The microclimates create many ecotones (areas of

transition between two different plant communities), and today farmers are exploiting these niches to grow an ever stronger portfolio of crops.

In Western Washington, the landscape at the time of first white settlement was almost entirely covered with coniferous forest. Sitka spruce dominated the coastal forest; to the east and at higher elevations, western hemlock was the dominant species. In the Puget Lowlands, the dominant species was Douglas fir, which has the widest distribution of the native forest species on both the west and east sides of the Cascade Mountains. At higher elevations, Douglas fir or western hemlock forests give way to those dominated by Pacific silver firs. Alpine associations are found at elevations above about 4,000 feet near the Canadian border and 5,500 feet near the Oregon border.

The continuous forest was broken in Western Washington only in areas where effective precipitation was low—either on rocky, south-facing slopes in rain shadow areas (as in the San Juan Islands) or in glacial outwash plains (such as the Mima mounds). In these exceptional areas, grasslands replaced the continuous forest.

East of the Cascades, vegetation changes quickly as rainfall levels fall and as the climate becomes more continental. East-slope Cascade forests are more diverse than west-slope forests (which tend to be monocultural). A mix of pine, tamarack, fir, and deciduous species are found in east-slope forests. These forests become lower in biomass per acre as the climate becomes more arid, and gradually the forest is supplanted by a shrub-grassland community in the desert near the Columbia River. The desert vegetation community is dominated by sagebrush, while in the Palouse the natural vegetation is a shrub-steppe province mixture of grasses and shrubs.

Influence of the Environment on the Seattle Population and Economy

At the global scale, Seattle's dominant influences are (1) its position at the northwestern extreme of the United States, which favors trade and other relations with Alaska and Asia; (2) its temperate marine position on the Pacific and at latitude 47. The first influence early on created opportunities for and even dependence on trade. The latter, together with the landforms discussed above, supported the immensely high-quality forests, and also fisheries, that were the basis for much of the population and economic growth of Seattle's first century.

At a regional scale, the relationship of people to the environment is complex. Over the lowlands are areas for which agriculture is suitable and can be profitable and much larger areas for which forestry remains viable, though both are impacted by urban and industrial development. The mountains include large areas of working forests but

also have many exceptional scenic areas, some of which have been recognized and protected as national parks and wilderness areas. Less spectacular but more pervasive are extensive networks of lakes and streams, popular for recreation and for urban residential and recreational housing. The impressively large mileage of Puget Sound waterfront is both a coveted resource for development of every kind and increasingly a target for protection and restoration, and thus an environment of continuing controversy. Puget Sound's several natural harbors have been quite favorable to port development.

At the more local scale, the microenvironment again creates opportunities for and imposes constraints on human use. The hilly and lake-studded terrain in urban areas puts a premium on waterfront and view properties, which differentiates areas by class. Seattle's interspersed land and water bodies, with steep slopes and wetlands, mean that the physical geography clearly channeled development in particular areas. It also required a number of costly investments to expedite movement, including the Denny Regrade (see chapter 6) and its numerous bridges, including two over Lake Washington and six over the Lake Union ship canal.

Conflicts over the use of land have a long history. Thus, the filling of miles of wetlands for industrial and commercial development is now giving way to preservation and even some restoration. Areas like the Green River Valley south from Lake Washington (Renton) to Auburn are classic examples of relatively rich farmlands yielding to transcontinental railroads, shopping centers and industry, but also of efforts to maintain remaining farmlands and wetlands.

INTRODUCING GEOGRAPHY

Michael Brown

"Oh, yeah, well, what's the capital of Maine?"[1]
"Geography! Like on the game show *Jeopardy*?"[2]
"Where's Timbuktu, then?"[3]
"Yeah, is that about rocks and stuff?"[4]
"Dude, don't we know where everything is already?"[5]

These are some responses you can get when you tell people that you are a geographer. And while we tend to leave rocks and stuff to geologists, all of these sound bites are frustrating because they are, indeed, part of our intellectual heritage. But geographers are so much more! Trying to pin down what this discipline called "geography" actually is can be no less frustrating. Before we go much further, I should say something about the discipline. Some of you might know geography in terms of its five themes: location, place, human-environment relations, movement, and region. Others may recall it as a one-off subject around seventh grade. Here are some useful definitions:

> "Human geography is the study of the spatial organiza-
> tion of human activity and of people's relationships
> with their environments."[5]
> "Geography is the why of where."[6]
> "Geography is to space as history is to time."[7]
> "Geography is what geographers do."[8]

While there have been geographers since ancient times ("geography" translates from ancient Greek as "earth writing" or "earth description"), and in all societies, the emergence of modern academic geography really begins in the nineteenth century amid the formation of academic disciplines across the social sciences in Europe.[9] Geography has inevitably been a hybrid discipline, splitting—and often connecting—human and physical realms. Here at the University of Washington, geography is located in the division of social sciences, rather than the physical sciences (see Appendix 1). The philosopher Immanuel Kant claimed that geography and history were unique among the social sciences as they focused on dimensions of existence (geography: space as history: time), rather than on components of social life (economists study the economy, political scientists study politics, etc.).

And, like many of its fellow social sciences, in a geopolitical context, geography was historically in long-standing service to imperialism.[10] It explored unknown and unmapped parts of the world. It inventoried potential and conquered colonial territories for valuable natural resources to extract and valuable labor to exploit. It produced information about indigenous societies and cultures in order to make them governable and productive. It aided military and diplomatic branches of government in strategies of war and international relations. By the early twentieth century, academic geography was typified by the regional approach, wherein the earth was broken up into distinct, cogent units, and scholarship demonstrated that this classification produced functional wholes and integrated, synthetic knowledge about these regions.[11] Regions can also describe the internal organization of nation-states. Chapter 4 in particular reflects this legacy in part.

Throughout the mid-twentieth century, this regional approach caused geography to lose respect, and it came under threat by an increasing academic division of labor. In the late 1940s, the president of Harvard University reviewed his geography department and found its largely descriptive and cataloguing approach to be "remedial" and not worthy of a university discipline. He closed it, setting off a chain of geography department closures in the United States that nearly caused the demise of the discipline.[12]

What saved geography was the "Quantitative Revolution" of the late 1950s. Young geographers had been trained during World War II to use quantitative methods of analysis, theories of probability and inference, as well as new computing equipment to handle quantitative data in their academic careers. More importantly, those geographers then trained graduate students (at UW they were known as the "space cadets") in these techniques and in the history and philosophy of science. Geography turned away from "mere" empirical description and toward deductive hypothesis testing, probabilistic modeling, and the search for universal laws. The Geography Department at the University of Washington was a key node of this new approach, when geography became a science of space.[13]

The revolution reinvigorated the discipline and garnered it interdisciplinary respect from other social sciences. Geography's scientific basis produced new sub-disciplines, such as spatial analysis, but it also changed the orientations of the older systematic branches, such as economic geography (see chapter 2), political geography (electoral geography remains quite quantitative, for example), and social geography (see chapter 6 on the housing crisis).[14]

After the late 1960s, another revolution of sorts again reoriented the discipline. Several geographers were disillusioned with the lack of political saliency and effective social change in their field. They were witnessing massive economic, political, and cultural shifts in the places they studied and were frustrated by what they saw as the naive objectivity that some social scientists displayed in the face of stunning violence, inequalities, and oppressions. While one of the developments of scientific geography had been the introduction of theory into geography, some geographers were frustrated with its lack of depth or help in explaining spatial patterns that they were exploring with ever-increasing methodological complexity. Better explanations and understandings of the world and its problems were needed if geographers were to play a role in solving them. Some geographers turned toward Marxism—without necessarily abandoning their scientific approaches or techniques—as a means of explaining the causes and consequences of capitalism throughout economic, political, and cultural histories and geographies. Marxism's rich intellectual depth and insights into the cumulative and seemingly disconnected elements of social inequality gave geographers powerful analytical tools not just to critique but to explain social injustice.[15]

This theoretical turn has flourished into a dizzying array of intellectual trajectories, as what counts as a geographic topic has greatly expanded. Over the past twenty-five years, there has been an explosion of interest in all manner of critical and radical social theory: humanism, realism, feminism, critical race theory, queer theory, postmodernism, poststructuralism, actor-network theory, and nonrepresentational theory are among some of the more familiar "isms" that have broadened the intellectual paths geographers have taken in thinking about how the world works (or doesn't, as the case may be).[16] Empirically, geographers study all parts of the globe, but they also focus their inquiry at all spatial scales: the globe, the nation, the state, the region, the locale, the neighborhood, the home—even the body. Indeed, some of the most fascinating work is that which juxtaposes or tacks between the local and the global, as you'll see in the following chapters.

Furthermore, it would be erroneous to assume that these changes have meant the demise of scientific or quantitative geography. These approaches themselves have made great advances in methodological and analytical sophistication since the early days of the revolution. The advent of geographic information science (GIS) has not merely "computerized" pen-and-ink mapmaking (cartography), it has allowed us to visualize and represent more and more complex fields of data, and it has been a key force in showing the power of geography. Indeed, some of the most interesting work in the field currently melds complex methodological and GIS elements with critical social theory (see chapters 5 and 6).

THE DEPARTMENT OF GEOGRAPHY AT THE UNIVERSITY OF WASHINGTON

Richard Morrill and Michael Brown

The history of the Department of Geography at the University of Washington reflects the broad trends summarized in the previous section, though some trends were more "placed" here than others. All of the authors in this volume are (or have been) faculty members and students of the department. Although there were professors of geography from the 1920s, the UW department as such dates from its separation from Geology (now part of Earth and Space Sciences) in 1935. For over fifteen years, the program was built and guided by the economic geographer Howard Martin. Climatology (now Atmospheric Sciences) was separated in 1947. By the late 1940s and early 1950s, the department had developed a stellar reputation in most of the specialties that remain sixty years later—economic,

1.8   University of Washington with Mount Rainier in the distance.

political, and urban geography, cartography, human-environment relations, and international studies (including Asia, Latin America, and Canada).

The "revolutionary" era really began in 1950 with the hiring of Edward Ullman and William Garrison, and it was supported by Marion Marts and the chairman Donald Hudson, also brought in in 1950. The so-called quantitative turn began in 1954, with the infamous "Geography 426" (a quantitative techniques course that is still taught in the department), and burgeoned in 1955 and 1956, with the arrival of a swarm of "space cadets" (including Richard Morrill). This tradition, which we prefer to view as a social science blend of theory and method, had a profound influence on geography as a discipline, which continues through teaching and research in economic, urban, social/population, medical, and regional specialties. These space cadets went on to train students, including current faculty like William Beyers and J. W. Harrington (whose work appears in chap-

ter 2). And the quantitative tradition lives on in the work of geographers like Kam Wing Chan, Suzanne Davies Withers, and Mark Ellis.

Current departmental strengths include political economies of globalization, global health and poverty, Geographic Information Science (GIS)—especially concerned with questions of decision making and community development, both urban and rural geographies, migration and population studies, and political geographies of various sorts—and regional expertise in Africa, Asia, Europe, Latin America, Russia, and North America. Work in the department is widely informed by the array of critical and radical social theory, reflecting the trajectory of the discipline discussed above. We study phenomena at the global as well as the micro scale. While we study a wide variety of issues from a variety of intellectual and methodological perspectives, it has often been said that we all do theoretically informed empirical work. In other words, we engage the real world and its problems but also take careful consideration of how to *think better* about them.

A remarkable feature of the department has always been its collegial atmosphere, and high level of communication, in support of all students and faculty, welcoming and blending humanistic, cultural, and social theoretic approaches. The department was in the vanguard of fostering and supporting women students and faculty and has long been welcoming of GLBTQ faculty and students. There has always been a strong sense of caring and compassion in the department. We may be almost unique in the remarkable loyalty and continuity of the faculty. A full list of faculty can be found in Appendix 2.

AN INTRODUCTION TO THE BOOK

Michael Brown

This book represents the collective efforts of faculty, graduate students, and undergraduate students in the University of Washington's Department of Geography at the end of the first decade of the twenty-first century. We came together to produce this volume on the occasion of the 75th anniversary of the department. Each year, the Association of American Geographers (AAG) hosts a conference in which geographers and their fellow travelers gather in a different North American city to exchange and disseminate cutting-edge research and to collaborate and inspire one another. The 2011 meeting of the AAG is scheduled to be held in Seattle. The last time the meetings were in Seattle was 1974, when the department put together a volume similar to this one.[17] So, a key purpose of this book is to introduce our geographic colleagues to our city and our region, and to their links with other places around the globe. We hope that conference attendees will find our lessons and insights about Seattle compelling enough to stir their geographical imaginations.

But we never intended our audience to be only other professionals. Two central missions of the UW Geography Department are "accountability to place" and "public scholarship." Accountability to place means that we feel a deep responsibility to the place in which we live. It is our responsibility to help make it better, to expose its problems and inequalities, to suggest solutions. We are held accountable to use our skills as teachers and researchers, as geographers. We care for our place. While not all of us study Seattle per se (or at least not all the time), this volume embodies a collective moment in which we exercise that care and attentiveness to Seattle—and the Northwest more generally—as a place.

"Public scholarship" is a term very much in vogue at the moment. It refers to the need for academics—especially those at a top research institution—to step away from solipsism, to pause their discussions and debates with other academics, to suspend their use of opaque jargon and speak directly to the community and the public.[18] Higher education in the United States—and here in Washington—is coming under increasing criticism and scrutiny about its relationship to the geographies it serves, especially in large research institutions like the University of Washington. Critics in Seattle often decry the UW as being walled off from the city and region (falsely, we think), taking ever more money and space and giving back precious little. This book is our modest effort to counter such misconceptions. We cannot speak for the entire university—or even for other departments, though one might extrapolate from us to them—but we are committed to showing you some of the ways in which our commitment to place is manifest in both our teaching and our research.

When we decided to edit this book, we deliberately gave potential authors a vague brief: we asked them to contribute short pieces that relate to their research—and their teaching—on Seattle and its region. We purposefully did not allocate chapters or topics, because we wanted this volume to be about their interests and passions. We did not assign people to collaborative writing groups in advance; contributors either made their own chapter groups (e.g., chapter 4) or we combined related segments into chapters (e.g., chapters 2, 5, and 6). The result is a text that reflects both regional points of focus and standard systematic formats (economic, political, social, and cultural). This combination reflects the simple truth that while we are all of a single discipline, how we situate ourselves and each other therein varies and is contested.

The book is organized systematically, reflecting common divisions of labor in human geography. Economic geography is discussed in chapter 2, political geography in chapter 5, social geography in chapter 6, and cultural geography in chapter 7. Not all of us approach our research so systematically, however, and so rather more of a regional or chorological focus is adopted in chapter 3, which situates Seattle in the currents of contemporary globalization, and chapter 4, which looks beyond the city of Seattle toward the wide expanse of the rural Northwest.

In chapter 2, economic geography is covered both widely and in depth. William Beyers begins with a historical geography of the region's economic development. Richard Morrill focuses on the contemporary geographies of employment. The key processes that produce local entrepreneurship here in Seattle are explored by J. W. Harrington and Charles Kauffman, while Nicholas Velluzzi offers a similar analysis for the Walla Walla Valley. Kam Wing Chan and Spencer Cohen detail the Northwest region's trade relations with China. Finally, William Beyers narrates his fifty-year career as an economic geographer who has been deeply involved in local issues.

In chapter 3, Matthew Sparke offers critical ruminations on Seattle's status as a "global city." He situates Seattle in broader networks of competition and collaboration, as well as hones in on the city's leadership role in global health initiatives and dilemmas. His chapter exemplifies a fundamental geographic point: In order to understand a place, one often has to look more broadly than just within its borders. Part of what makes Seattle a vibrant and compelling city is its situation in broader networks of power, capital, and knowledge.

Continuing with the theme of geographic breadth, David Barker and colleagues reorient our focus away from the city per se and toward a broader perspective on the rural Northwest in chapter 4. Their careful analysis of layered old and new rural geographies tracks the profound economic, demographic, and cultural restructurings that

are shaping nonurban areas. They offer images of both "playgrounds" and "dumping grounds" in these new rural spaces: places of in-migrating wealth alongside poverty and a rising prison population.

Chapter 5 is a comprehensive introduction to the different forms of the area's political geography, along with an analysis of its electoral geographies, by Larry Knopp and Richard Morrill. Steve Herbert offers a reflective piece on Seattle's political culture, specifically, its paradoxical attitude toward social control of marginalized groups. Who controls public space, and how? John Carr investigates that same question in the context of disputes between youths and elders about the emerging hipster landscapes of skate parks in Seattle. Timothy Nyerges, Kevin Ramsey, and Matthew W. Wilson tackle one of the most entrenched issues on the local political scene—transportation—through a research project that used GIS experimentally to make the decision-making process more democratic and efficient. Kevin Ramsey offers a short piece on the way debates over climate change and the notorious Alaska Way Viaduct were framed, and Matthew Wilson reports on efforts to empower neighborhood residents, using GIS technology to document and report problems more quickly. Finally, Sarah Elwood discusses political geography of the area in terms of one of its rapidly growing but difficult to see characteristics: the rise of the voluntary sector.

Chapter 6, "Social Geographies," is the longest chapter, with the greatest number of authors. It focuses on four broad areas: population, housing, education, and neighborhoods. Richard Morrill provides maps and data for a population geography of the area. Suzanne Davies Withers looks at the recent mortgage crisis and demonstrates the surprising variability of effects geographically. Tony Sparks tackles another type of housing crisis—homelessness. From his ethnographic research, he takes us to Tent City 3 for a view inside a homeless community.

Turning to educational issues, Tricia Ruiz and Mark Ellis analyze demographic data about assignment to and enrollment in the Seattle Public Schools. Their results show the racial resegregation of the system, despite more than

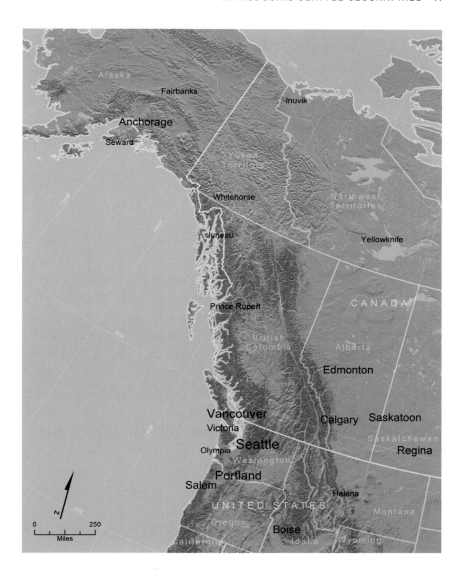

1.9   Regional map of Seattle's situation in the Pacific Northwest. Source: ESRI Data and Maps 9.3.

three decades of efforts to combat that spatial process. Busing was the major policy initiative to desegregate the schools here, and Catherine Veninga reports on her novel research with former bused public school students to determine if that controversial policy was successful. Gentrification is addressed by Kim England and Gary Simonson. England provides an historical account of Belltown, Denny Hill, and the Pike Place Market to show how questions of access to and control of urban neighborhoods are recurrent themes over time. In a very different part of the city, Columbia City, Simonson interviews long-term residents

("stayers") about the effects of rapid gentrification of their neighborhood. Finally, Michael Brown, Sean Wang, and Larry Knopp problematize the common wisdom that Capitol Hill is Seattle's gay neighborhood.

Cultural geographies are again addressed in chapter 7, written by students in the 2010 Geography Department Honors Program, under the supervision of Katharyne Mitchell. The students offer their observations on the cultural landscapes of the city: the region's complicated relationship with its natural symbol (and key commodity), the salmon; the shift from an industrial to a postindustrial city dominated by services and the information economy ("coffee and the microsofty"); grunge and the local music scene; how the alternative food movement is changing the Seattle landscape; and the cultural legacy of the WTO (World Trade Organization) riots. These essays reflect not just the geographies of the region but also the teaching and education functions of the department. Simply put: this is how a group of smart, committed students put what they learn into practice and make it relevant to themselves and their readers.

Because of this inductive approach, the book's content might read as uneven with respect to particular subfields or topics for which this department is known (health geography or feminist geography, for example). This is not an unusual problem in the discipline; since there can be a geography of everything, what do you leave out? How do you ever finish? Yet the strength of the book is that it captures this department's immediate, specific, and engaged research on and understanding of this place in which we live. Geography—this "earth writing"—is always uneven and incomplete. There is always more that can be said, but, for now, we invite you to explore Seattle and its region through geographers' eyes.

NOTES

1   The capital of Maine is Augusta.

2   Timbuktu is a city in the Tombouctou part of Mali, West Africa.

3   No, that would be geology, though some geomorphologists are interested in rocks to the extent that rocks cause and affect the shape and form of the Earth's surface.

4   To a large extent, yes, we do know where things are. We've mapped most of the planet, and the amount of geographic information has exploded in recent years. But, of course, we don't know the location of things we don't know about (yet). Also, the premise of the question is that location or spatial processes never change—but they do.

5   Paul Knox and Sallie Marston, *Human Geography* (New Jersey: Prentice Hall, 2004), 2.

6   Billie Lee Turner, personal communication, c. 1986, Clark University, Worcester, MA.

7   Stuart Elden, "Reassessing Kant's Geography," *Journal of Historical Geography* 35 (2009): 3–25.

8   Michael Zimmer, personal communication, c. 1986.

9   Geoffrey Martin and Preston James, *All Possible Worlds* (New York: John Wiley, 1993); David Livingstone, *The Geographical Tradition* (Oxford: Blackwell, 1992).

10 Neil Smith and Anne Marie Godleweska, *Geography and Empire* (Oxford: Blackwell, 1992); Felix Driver, *Geography Militant* (Oxford: Blackwell, 2001).

11 Susan Schulten, *The Geographical Imagination in America* (Chicago: University of Chicago Press, 2001).

12 See Neil Smith, "Academic War over the Field of Geography," *Annals of the Association of American Geographers* 77 (1987): 155–72.

13 Peter Gould and Forrest Pitts, eds. *Geographical Voices* (Syracuse, NY: Syracuse University Press, 2002).

14 R. J. Johnston and James Sidaway, *Geography and Geographers*, 2d ed. (London: Arnold, 2004).

15 Richard Peet, *Modern Geographic Thought* (Oxford: Blackwell, 1998).

16 Paul Cloke, Chis Philo, and David Sadler, *Approaching Human Geography* (New York: Guilford, 1992).

17 A. Phillip Andrus, William B. Beyers, Ronald Boyce, Jacob Eichenbaum, Michael Mandeville, Richard Morrill, David Stallings, and David Sucher, *Seattle* (Washington D.C.: Association of American Geographers, 1976).

18 Katharyne Mitchell, ed., *Practicing Public Scholarship* (Oxford: Blackwell, 2008).

TWO

ECONOMIC GEOGRAPHIES

William Beyers, Richard Morrill,
J. W. Harrington, Charles Kauffman,
Nicholas Velluzzi, Kam Wing Chan,
and Spencer Cohen

How has the local economy evolved? Where are different types of jobs in the metropolitan area located? How does a region reinvent its economy or trade successfully? We answer these questions in this chapter. Economic geography studies all types of economic activity at a variety of spatial scales. Here we offer several different perspectives on the economic geography of the Seattle region: historical, evolutionary, and contemporary. We highlight not only the emergence of specific forms of extraction, production, and consumption, but also the ways in which spatial interactions between Seattle and elsewhere have buoyed the local economy. Trade with China is a good example of this interaction. We show the importance of place in bringing key factors together to produce entrepreneurial success in both Seattle and an emerging regional economy in Eastern Washington.

ECONOMIC GEOGRAPHY OF THE REGION OVER TIME

William Beyers

....

The development of Seattle's economic geography can be understood as a unique combination of "exogenic" (external to the area) and "endogenic" (internal to the area) forces.[1] Prior to 1851, there was no permanent white settlement in Seattle. Early explorers (such as George Vancouver in 1792) identified the geography of Puget Sound and gave it many of the place names that we use today. Later, fur traders spread through the region, with a small settlement at Fort Nisqually serving as a trading post for these fur traders. However, the "initial point of attachment" occurred on September 28, 1851, when the Denny party landed and set up camp at Alki in what is now West Seattle. In December

1851, this group of early settlers was visited by a sailing vessel from San Francisco and asked to supply logs to be taken to San Francisco for processing into lumber to fuel the growth of the Bay Area in the wake of the California Gold Rush.

The Alki location was very exposed to strong north winds, and so, in March 1852, most of the Denny party moved their settlement from Alki to what is now downtown Seattle. Henry Yesler moved into this new settlement in 1852, and in October of that year, he started his sawmill, which began to provide lumber to ships calling from San Francisco. Within a few years other mills were built to process logs into lumber, and multiplier and structural forces took effect, stimulating settlement in what is today the Puget Sound region.

Early History: 1851 to 1897

The log trade quickly turned into a lumber trade, since lumber is lighter to ship than logs. In the early pre-railroad era, logs were skidded into Puget Sound and towed to lumber mills for processing. By the late 1860s, logging railroads began to be built to access more distant lowland timber supplies, but prior to the arrival of the transcontinental railroad in 1883, all exports of lumber went by sailing vessel, primarily to California markets.

Puget Sound was a frontier outpost, dependent upon the import of commodities such as clothing, liquor, and the machinery used in the lumber trade. Seattle itself emerged as an early center for trading these imported commodities and for producing services to meet regional demand. The early export trade stimulated the production of other goods and services in the local area. Farms, retail outlets, and schools of higher education followed quickly. Maynard, Boren, and Bell, for example, became leading retail merchants, and the University of Washington was founded in 1861 to provide educational services to the region. Agriculture began shortly after initial settlement, providing vegetables, fruit, and dairy products. Fishery resources were also tapped for foodstuff, but the peak level of commercial fishing activity in Puget

Sound did not take place until after 1900. Mining also began in the early days. Coal discoveries in King County led to the construction of rail lines to Renton and Newcastle, and coal was brought to Seattle for export.

Construction of the transcontinental railroads to the Seattle area was completed in 1883, with the arrival of the Northern Pacific line in Tacoma. There was significant rivalry among towns in the Puget Sound region for the terminus of the Northern Pacific and other proposed rail lines into the region. In 1884 the Northern Pacific route was extended to Seattle, and in 1893 the Great Northern Railway completed its tracks into the city. The Northern Pacific was a land-grant railroad, receiving 11 million acres of land in its corridor from the Midwest to Puget Sound.[2] It became a significant force in real estate development through the disposition of most of its land grant after construction of its route, although a sizeable portion of this land grant remained in Northern Pacific hands in the Cascade Mountains until modern times. In contrast, the Great Northern Railway did not receive a land grant and was dependent upon commercial traffic to make a profit.

By the mid-1890s, the Great Northern had captured a large share of the eastbound lumber trade, which in turn boomed in Western Washington with the arrival of the railroads. The region was no longer solely dependent on water-borne transport to ship goods to export markets, and the cost of movement by land was much lower than by water. The Northern Pacific Railway was forced into receivership by the Great Northern in the mid-1890s. Great Northern tried to take over the Northern Pacific in this time period, but the courts blocked the merger on anti-trust grounds. To regain solvency, the Northern Pacific sold its land grant in Western Washington to the Weyerhaeuser Company, a Minnesota-based timber concern, which immediately became the largest lumber company in the region.

From the Alaska Gold Rush through World War I

On July 17, 1897, a ship from Alaska landed in Seattle with gold. Word spread immediately about the opportunities to strike it rich in Alaska and the Yukon, precipitating a huge migration of fortune seekers. Seattle's population swelled from 80,671 in 1900 to 237,194 in 1910, and to 315,312 by 1920. Most prospective miners never went to Alaska, but instead worked in the natural resource–based sectors in the region—timber, fish, agriculture, and mining. Seattle grew as a center for wholesaling, the focal point for the distribution of goods into markets in Western Washington, as well as the center of brokeraging exports into domestic markets by the railroads and abroad through maritime commerce.

The railroad construction era brought Asian (mostly Chinese and Japanese) migrants to the region; in the wake of the railroad construction boom, many of these newcomers became engaged with the timber industry and agriculture. Seattle's International District traces its roots to these late nineteenth-century immigrants.

Timber production in Washington State continued to swell after 1900, reaching an all-time peak in 1929.[3] Corporations such as Weyerhaeuser solidified their power as timberland owners and diversified processers of timber products. Salmon harvests boomed in Puget Sound after the turn of the century, peaking in 1910. Agriculture continued to spread through lowland valleys, especially in areas cleared of western red cedar where soil quality was high.

World War I brought a boom in naval activity to the region, and the start of the Boeing Company, whose founders, Bill Boeing and Conrad Westervelt, landed a contract to build airplanes for the navy during World War I. The war was over before they were able to produce and deliver these airplanes. After World War I, the company turned to furniture manufacture for a while, but pursued innovations in aircraft design in the 1920s, and pioneered aircraft carrying mail. Boeing executives realized that if their planes could carry mail, they could carry passengers too, and in 1929 they started a passenger airline (now United Airlines). At the same time, Boeing also went on an extensive acquisition program, including purchase of what is today Pratt & Whitney.[4] Boeing was forced by anti-trust laws to divest itself of many of these vertically integrated businesses in the late 1930s.

The continued development of natural resource–based industry, plus the growth of industries important in World War I, pushed up population in the Greater Seattle area from 482,000 in 1910 to 634,000 in 1920, and 738,000 in 1930. The Great Depression years saw a stabilizing of the population and some movement back to the land.

World War II through the Vietnam War Era

World War II brought the Puget Sound region out of the doldrums of the Great Depression. The Boeing Company assumed a central role in the war effort for airplane manufacture; employment swelled at the firm from 4,000 in 1940 to about 50,000 in 1944.[5] Naval activity in the region was also strong, with local shipyards turning out vessels; the Bremerton Naval Shipyard played a key role in the repair of navy vessels, employing 32,500 people in 1945.[6] The region became a key support location for the military, with bases in Kitsap and Pierce counties. After World War II, Boeing employment fell sharply, but the Cold War and the advent of a new generation of aircraft based on jet engines quickly led to resumed job growth. Aircraft such as the B-47 and B-52 were developed for military use in the Korean War and elsewhere, and Boeing adapted these designs to passenger jets with the development of the 707 in 1957.[7]

The surge of business in aerospace and military activity helped spark growth in services in the region, such as the Port of Seattle, wholesaling, rail transportation, and related logistical services. The advent of container cargo handling, the recovery of Asian economies, and their nascent development as export platforms (especially Japan, South Korea, Taiwan, and Hong Kong) were all related to new beginnings in international trade. Puget Sound also remained a key location for trade with Alaska in the post–World War II era. The Vietnam War brought the region again into focus as a support location for military activity.

In the period from 1940 to the Korean War in 1950, population in the region swelled from 820,000 to nearly 1.2 million, a gain of 46 percent. This was the largest absolute gain in population in the region's history, and it was followed by continued population growth in the 1950s and 1960s. By 1960 regional population was 1.5 million, a gain of 26 percent from 1950. Table 2.1 provides employment data from 1940 to the present, divided into broad categories.

The post–World War II era was also partially driven economically by the clout of the Washington State congressional delegation, the rise of the local environmental movement, and the development of a number of significant cultural institutions. The legacy of this period of development was more vivid in later years, but the seeds of change were being planted. U.S. Senators Warren G. Magnuson and Henry M. Jackson helped push through Congress key environmental laws, such as the Coastal Zone Management Act, the National Environmental Policy Act, the Wilderness Act, the Clean Air and Clean Water acts, and legislation specifically targeted at land preservation in the North Cascades, including North Cascades National Park. The populist political atmosphere of the region helped to spawn powerful local environmental organizations, such as the Mountaineers, the North Cascades Conservation Council, and the retailer REI. These initiatives helped give the region a reputation as "green" and were likely instrumental in attracting economic actors enamored with these values.

Sir Thomas Beecham, conductor of the Seattle Symphony (1941–43), referred to the region as an "aesthetic dustbin," when he arrived in Seattle in 1941.[8] However, artists such as Mark Tobey, Kenneth Callahan, Morris Graves, and George Tsutakawa drew on the lush environment to create the Northwest School of art that became internationally recognized in the post-war era. The University of Washington Schools of Art and Music began to have a strong impact on local artistic institutions, and the Seattle Art Museum and the Seattle Symphony had sustained growth in the 1950s and 1960s.

The arrival in 1958 of Charles Odegaard as president of the University of Washington marked a change in the atmosphere of the university, with a much stronger emphasis on research. With the support of powerful Senator Warren Magnuson, chair of the Senate Appropriations Committee, the University of Washington Health Sciences programs benefited from sizeable increases in federal research support. By 1974 the University of Washington rose to have the highest

TABLE 2.1 EMPLOYMENT IN THE PUGET SOUND REGION, 1940-2007

	Primary	Mfg & Construction	Wholesale & Retail	Other Service	Government	Total
1940	19,870	87,486	64,811	101,269	21,064	294,500
1950	18,843	125,400	97,021	149,661	27,744	418,669
1960	12,784	182,823	109,022	191,400	28,496	524,525
1970	5,858	126,358	119,923	199,965	95,293	547,397
1980	12,317	184,549	187,342	319,951	111,473	815,632
1990	17,300	240,930	257,646	518,366	137,815	1,172,057
2000	20,504	232,042	316,661	716,337	162,547	1,448,091
2007	15,885	369,980	332,279	1,324,732	342,090	2,384,966

NOTE: Discontinuities between early Census data and Bureau of Economic Analysis 1970–2000 and 2007 are due to classification differences. These data are inclusive of self-employed persons plus wage and salary employment.

SOURCE: 1940–1960 Census of Population, 1970–2007 BEA, Local Area Personal Income Series SA-25.

level of federal research funding of any public university in the United States, a position that it has kept since that date. Clearly, this was an era in which the place of the university in the fabric of the local economy expanded dramatically.

The Modern Era: About 1970 to the Present

The Puget Sound region in the period from the Vietnam War to the present has been on yet another economic trajectory. The seeds of the modern era were sown in World War II, but the region has had a blossoming of new industries in this recent era. The key sectors identified here are (1) technology-based industry; (2) the ports; (3) producer services; (4) arts, sports, travel, conventions, cruise ships; (5) health care; and (6) the military. These sectors are considered "key," as they have a significant trade dimension.[9]

Hi-tech or technology-based industry has become critically important to the Puget Sound region's economy in recent decades. These are industries that utilize relatively large numbers of employees in scientific, computing, and engineering occupations. Technology-based industry accounted for 40 percent of employment in Washington State in 2009. Most of this high-tech employment is located in the central Puget Sound region, while Vancouver, Spokane, and the Tri-Cities have smaller concentrations of employment in these industries.

Historically, manufacturing dominated high-tech employment in this region, and, in turn, aerospace (Boeing) was dominant within the manufacturing sector. Figure 2.1 shows the level of high-tech employment in Washington State from 1974 through 2007. This figure indicates that aerospace accounted for about half of high-tech back in 1974, but by 2007 this share had fallen to 23 percent. Figure 2.2 shows the mix of high-tech industry in Washington State in 2007, with manufacturing accounting for one-third of this employment and services accounting for two-thirds.[10] Gradually, the share of high-tech employment in Washington State and the Puget Sound region has shifted toward service industries, as depicted in figure 2.1.

Software and computer services must be singled out

as one of the most dynamic components of the economy in the last twenty-five years. Dominated by Microsoft, the local computer services sector has become second only to aerospace as a source of job creation in the local economy. However, jobs are not the only measure of impact, and it is very clear that "Microsoft money" has had a very influential spinout impact. The business ventures of Paul Allen, co-founder of Microsoft with Bill Gates, have been remaking the economy, including the development of a growing biotechnology cluster in the South Lake Union district and the Experience Music Project museum at Seattle Center. Thousands of Microsoft millionaires have invested their fortunes in business developments locally, such as Scott Oki's golf courses and Ida Cole's remodeling of the Paramount Theatre. Perhaps the most significant of these developments is the creation of the Gates Foundation, now the largest philanthropy in the world, with a strong focus on global health and education. It should be noted that this foundation includes not only the fortune of Bill and Melinda Gates but also that of Warren Buffett. Today this foundation has made large grants to start programs in global health and health-metrics at the University of Washington.

While the aggregate size of high-tech has expanded in the Central Puget Sound region, Boeing has been through various cycles of employment, as the demand for airplanes has expanded or collapsed. Figure 2.3 shows employment in Boeing/aerospace in the region from 1955 to 2009. Employment exploded in the 1960s, and then plummeted. By 1972 employment had fallen from the peak of 105,000 in 1969 to 39,000. Accompanied by high unemployment levels in other businesses locally, there were billboards in the early 1970s that famously asked, "Will the last person leaving Seattle please turn out the lights?" Aerospace employment recovered in the 1970s, reaching another peak in the early 1980s, followed by another slump and then a recovery, with yet another peak in 1990. Another slump in the early 1990s was succeeded by a buildup peaking in 2000, and then another drop, only to be followed by a weaker upturn over the last few years. In 2010 Boeing is laying off workers, due in part to the recession and in part

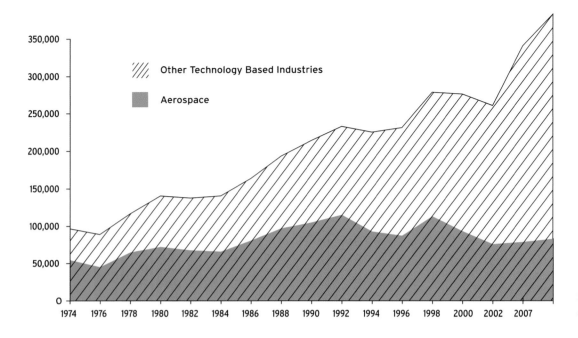

2.1 History of high-tech employment, 1974–2007.

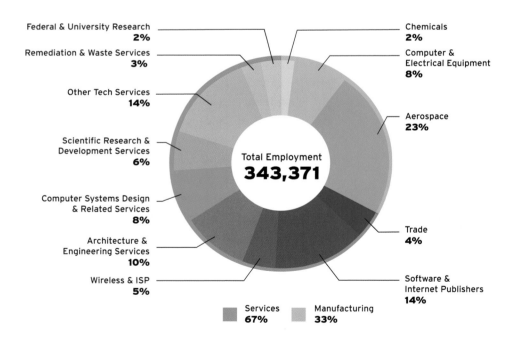

2.2 Distribution of high-tech employment, 2009. Source: Washington State Department of Employment Security.

to improved productivity from technological changes. The latter factor is well illustrated by the new 787 product line, for which the company has outsourced ever more of the production process (as illustrated in fig. 2.4).[11] However, there have been difficulties with this highly decentralized production chain, and Boeing has responded by acquiring some of the suppliers in this assembly system and bringing them into the Boeing Company production system.

2.3 History of Boeing employment, 1955-2009.

While Boeing and Microsoft are the two most visible components of high-tech in the Seattle area, it should be emphasized that there are actually thousands of businesses in this sector. Most high-tech service businesses are small, and the research activity at the University of Washington (included as a part of the high-tech sector) is dispersed among thousands of recipients of grants and contracts. University research is dominated by health sciences funding, where there are close ties to the Fred Hutchinson Cancer Research Center, one of the leading nonprofit health research organizations in the world.

Emerging within the high-tech sector locally is biotech—a difficult to define and ever changing mosaic of companies and organizations. Centered in the axis between the University of Washington, Lake Union, and the downtown waterfront, this complex has a mix of research and production organizations, including pharmaceuticals and bioengineering and medical equipment. Many of these organizations have funding linked to the Gates Foundation and staff ties to the university or the Hutchinson Cancer Center.

Puget Sound is home to two of the key marine cargo ports on the West Coast—Tacoma and Seattle. These ports were early to convert to cargo containers, and they move more freight than any other West Coast ports except Los Angeles and Long Beach. They have worked to be fully integrated with the railroads, as most cargo entering and leaving Puget Sound ports is not destined for local consumption; rather, it is an export or import to distant hinterlands. In 2008 these key ports handled $75.3 billion in cargo, 76 percent of which were imports. In 2008 they moved 3.57 million TEU's (20-foot equivalent units or 20-foot cargo containers), of which 24 percent were associated with domestic maritime trade, 37 percent were full containers imported, 26 percent full containers exported, and 14 percent empty containers.[12]

Beyond the firms and industries included in high-tech, the Puget Sound region is a modest hub of finance and other producer-service activity. The region has never been a major banking center, and its largest indigenous bank—Washington Mutual—was swallowed up by Chase bank in the recent great recession, after it failed because of poor residential lending practices. The region has a strong portfolio of firms in law, insurance (e.g., Safeco), securities and investment (Frank Russell Company), architecture and engineering, consulting, accounting, advertising, computer systems design, and scientific research and development (such as PATH or the Hutchinson Cancer Center). Frank Russell deserves special mention; this historically Tacoma-based firm has now moved to Seattle into what was Washington Mutual's headquarters, and Russell's parent (Northwestern Mutual) has acquired this office building. The Russell Company pioneered consulting to pension funds—not owning their assets, but advising their corporate clients about investments. It was sold in 1999 to Northwestern Mutual; it is famous for the Russell 2000 and Russell 3000 stock indices, broad stock market indicators. It also sells specialized financial information for the myriad market areas in which clients need to be invested. Now Russell Investment, this firm epitomizes the niche market character of producer-service firms here—and elsewhere—playing to market opportunities through differentiation and specialization.[13]

E-commerce has had a strong start in the Seattle area. Amazon.com was founded here and has moved from being an online retailer of books to an online retailer of almost everything. Real Networks is now one of the largest online music subscription and downloading services, second only to Apple's i-Tunes. Companies like REI and Nordstrom have also established a significant online presence.

The central Puget Sound region has become one of the most military-dependent regions in the United States in the last half century. This complex is located largely in Kitsap and Pierce counties, with a smaller concentration in Snohomish County. Kitsap County is one of the most military-dependent counties in the United States, with both a large civilian and a large military workforce.[14] The civilian workforce is centered on the Bremerton Naval Shipyard, one of the navy's largest repair and maintenance bases. Thousands of enlisted navy personnel are housed in Kitsap County during ship overhauls. This county also hosts the Keyport Naval Warfare Research Center for research on torpedoes and sonar technologies. Related to the Keyport Naval Warfare Research Center is the University of Washington Applied Physics Laboratory, which undertakes research that is closely linked to

2.4   Dreamliner supply chain. *The Seattle Times*, 2005; reprinted with permission.

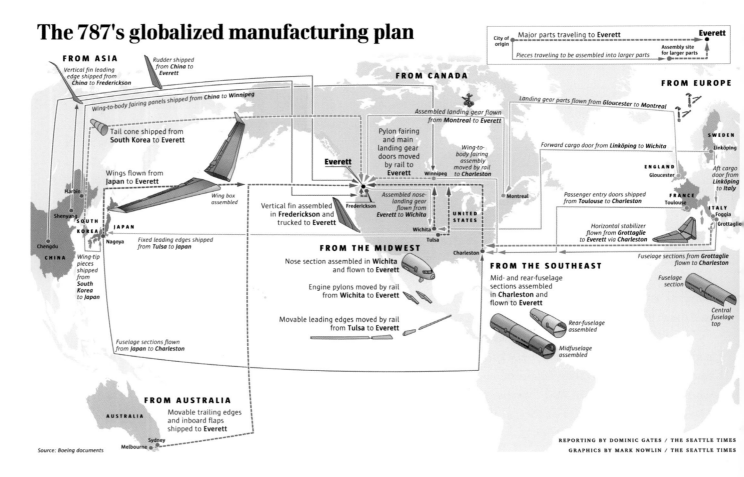

The 787's globalized manufacturing plan

the Keyport complex. Kitsap County also hosts the Bangor home base for Trident nuclear submarines. Pierce County hosts Fort Lewis Army Base and McChord Air Force Base, while Snohomish County has the new Naval Station in Everett, home port to several aircraft carriers.

Since Sir Thomas Beecham declared this region an aesthetic dustbin, things have turned around! Today, the central Puget Sound region is a thriving source of cultural and sports activities, as well as a major convention center. The region is the home of major league baseball, football, soccer, and women's basketball franchises, and the UW sports programs play an important role in the local sports scene (especially distinguished in crew over the years). The region boasts the third largest concentration of theaters in the United States, after New York and Chicago. Seattle Opera is noted internationally for its Wagner Ring Cycle, which in 2009 was sold out for three cycles, with people coming from forty-nine states and twenty-three countries.[15]

The Seattle Art Museum has opened the free Olympic Sculpture Park to the north of downtown. Benaroya Hall in downtown Seattle is not only filled with Seattle Symphony and chamber music events but also hosts performances by many other musical and cultural groups.

The city of Seattle has embraced music as a key economic activity, and it hosts the Mayor's Office of Film and Music. The broad extent of musical activity in the community is captured in figure 2.5, which traces the flow of activity from musicians to consumers.[16] The variety of modes in which this linkage can occur is demonstrated in the various streams in figure 2.5. These linkages may be local or tied to export markets. An estimated 22,000 people in the Seattle area are employed in the music sector, generating $4.6 billion annually in business activity.[17]

2.5 Seattle music industry production streams.

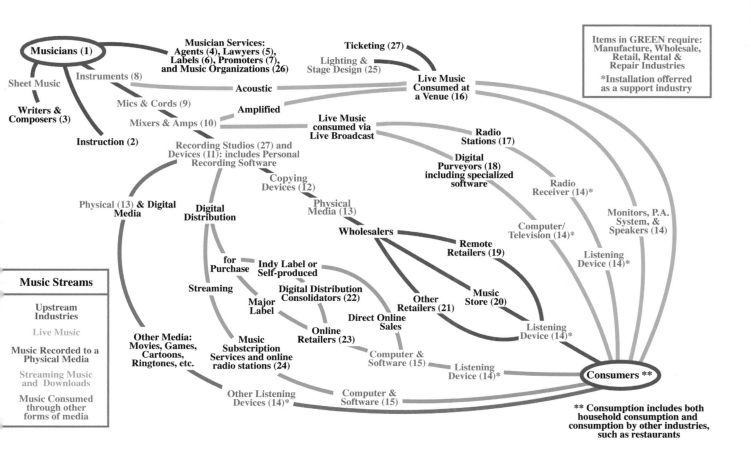

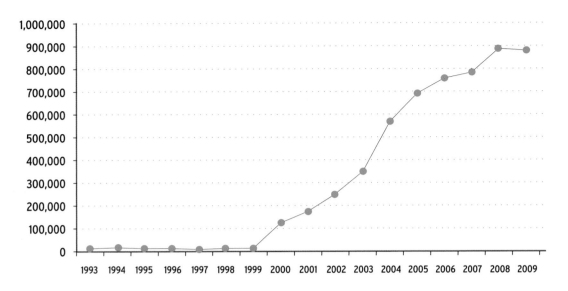

2.6 Cruise ship passengers, 1993–2009. Source: Port of Seattle.

The Seattle Convention and Visitors' Bureau estimates that King County had 9.34 million visitors in 2008, who spent $5.14 billion. Most of these people were on pleasure or vacation trips or were visiting friends and relatives. More than half arrived here by air, and over three-fourths were from U.S. origins. Just under half stayed in hotels or motels, with the balance staying in private homes and other venues.[18] The Port of Seattle has become a major location for cruise ship activity; figure 2.6 indicates that the level of cruise ship activity in this region has grown dramatically in the last decade.[19]

The Seattle area has a considerable health care sector, in both private and public employers. Major private hospitals are located in Seattle and Tacoma, while the University of Washington operates both University Hospital and Harborview Medical Center. The health care education and training activity at the university is linked with the research activity in health sciences and with research at organizations such as the Fred Hutchinson Cancer Research Center. Health-related activity is an important component of philanthropic organizations locally; the Gates Foundation, for example, has been instrumental in expanding the research focus of the University of Washington into fields of global health and health metrics.

GEOGRAPHIES OF EMPLOYMENT

Richard Morrill

Now that we have sketched the development of the region's economic geography, we turn next to elements of its contemporary makeup. Here we specifically explore the geography of the region's labor market. The distribution of jobs in relation to the location of population is shown in figure 2.7. The largest job concentrations are, first, in downtown Seattle and northward to the University of Washington and southward through the Duwamish valley. The second largest concentrations are clusters in Tacoma, Bremerton, and Everett (actually dominated by the Boeing plant in southwest Everett), but especially in the edge city of Bellevue, from downtown Bellevue northeasterly through the Microsoft complex in Redmond. The third largest concentration is in south King County, from Redmond through Sea-Tac, Kent, and Auburn, along the SR-99 and SR-167 corridors.

Reflecting these concentrations, most of Greater Seattle has far fewer jobs than people. The highest ratios of jobs to people occur from downtown Seattle south through the job-rich Duwamish corridor, then south through lowland Renton and Tukwila, Kent, and Auburn. High ratios

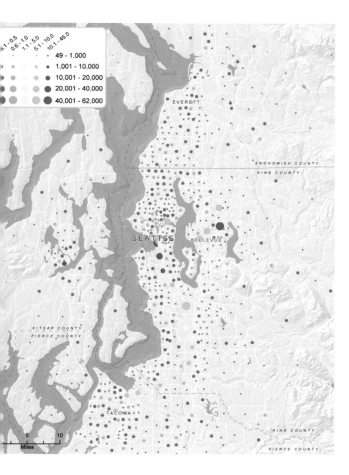

2.7 Distribution of jobs in 2005 census tracts.
Source: U.S. Census.

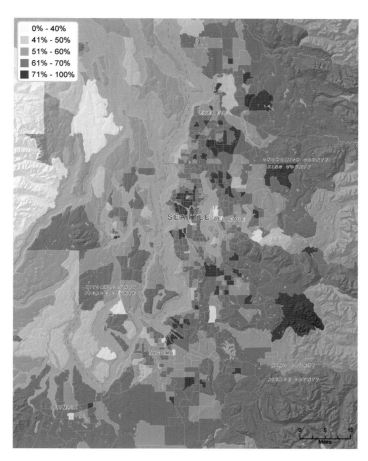

2.8 Female labor force participation in 2005.
Source: U.S. Census.

also occur in south Everett (Boeing) and the Port of Tacoma area, downtown Tacoma, and the corridor from downtown Bellevue through Redmond. This quite normal pattern of job-population imbalance defines the need for commuting, discussed in a later section.

Labor Force Participation and Unemployment

Male labor force participation has a distinctive pattern: high in most outer suburban areas, home to traditional families with children, and high in areas west from the university and northwest from downtown. Rates are lower (65–75 percent) in older suburban areas and very affluent areas, with higher shares of older populations, as well as in areas of high poverty and high minority shares (as southeast Seattle and beyond).

Female labor force participation (see fig. 2.8) is similarly high in areas west of the university and north of downtown, and in several middle and lower middle class suburbs, even among families with children but whose labor is needed. Participation is low in the poorest minority areas, with high shares of single parent families and/or of the foreign born, and also in the most affluent areas, such as Mercer Island.

The Distribution of Workers and of Jobs by Occupation and Industry

Managerial and professional occupations are associated with middle and upper classes, reflect the highest levels of education, and tend to occupy the most desired urban environments. Since they reflect higher incomes, they do similarly tend to locate reasonably near related job concentrations in FIRE (finance, insurance, and real estate); health, education, and government (higher level positions); and business services. These are most prevalent in or near downtown business districts and major health and education (university) complexes. For Seattle, location of managerial and professional occupations reveals an extreme concentration around Lake Union, from downtown Seattle to the University of Washington, and in Eastside communities, dominated by downtown Bellevue and the Microsoft campus in Redmond.

Clerical and office occupations are associated with middle classes, reflect intermediate levels of education (high school, some college), and are by far the most broadly, even somewhat evenly, distributed across the metropolis. Significant numbers of employees may be second wage earners in families. They are underrepresented only in the poorest areas, often with high shares of the foreign born. Service occupations, as defined by the Census, are associated with lower levels of education and locations of workers in the lowest income areas and those with the highest minority shares. They are much underrepresented in affluent Eastside areas.

Construction and crafts occupations are typically associated with the lower and middle class and intermediate levels of education. They are extremely underrepresented in the city of Seattle and in affluent Eastside communities, but are very common in extremely peripheral and even rural areas, in all counties of the region. This is in part related to a similar location of places of work and to the fact that many workers are independent contractors and work out of their homes. Production and laboring occupations are associated with the least education, usually high

school. In the Seattle region, these workers are strongly concentrated in the zone from southern Seattle through south King County, into much of Tacoma and central Pierce County, probably even more because of the location of affordable housing than because of the similar location of many of their employers.

The location of finance, insurance, and real estate jobs is extremely concentrated in downtown Seattle, followed by the downtowns of Bellevue and Tacoma, while the residences of FIRE workers are fairly widespread. While there are clusters in wealthier, often view, areas of central and north Seattle, the vast majority of FIRE workers reside in an arc from Bellevue and Mercer Island up the I-405 corridor to Snohomish County, with major clusters in Mill Creek and Edmonds. As expected, numbers are low in lower- and middle-class areas, especially in southern King and Pierce counties and generally in rural areas, even if affluent. Information services employ a subset of professional and managerial occupations. As expected, the distribution coincides with that of the highly educated, from north-central Seattle, the hinterland of the University of Washington, across Lake Washington, to and past Microsoft and other Eastside high-tech firms.

Retail trade is a widespread industry, but jobs in retailing are far more suburban than is generally realized. The largest numbers, and highest percentage of all jobs totally, pinpoint the region's largest malls and highway strip malls. The distribution of residences of retail workers is actually rather similarly located, although more diffuse around the malls and highway strips. Retail jobs are an especially high share of all jobs of residents in southwest Snohomish County, in central Kitsap, not in but north and south of downtown Bellevue.

Manufacturing jobs are quite concentrated in industrially zoned districts, south from downtown Seattle along the Duwamish waterway and major railways (the original Boeing plants, but a wide variety of manufactures), just beyond in Renton, with the Boeing 737 and Paccar (trucks) plants, then south through the Green River Valley (railroads, SR-167). The largest single concentration may be at the Boeing plant in southwest Everett. Other major clusters of jobs include high-tech firms to the east in Redmond and Eastgate, in the Port of Tacoma. As with retail, the residences of manufacturing workers closely reflect the clusters of jobs, and those in manufacturing are a fairly high percentage of all workers in most of Snohomish County and in south King County, especially in the Kent and Renton area. Shares are correspondingly low in central and north Seattle, where industrial operations tend to be converted into retail outlets.

The location of construction jobs reflects to some degree the site of major construction projects and temporary offices, as in downtown Seattle and downtown Bellevue, although the shares of all jobs are low. The larger concentration of construction firms is down the valley from Kent to Sumner, and in the Bothell area, while shares are rather high in far suburban and rural areas, where small independent contractors work out of their homes. The residential location of construction workers is extended to suburban and periphery rural in all counties of the Seattle region, suggesting that construction workers probably have the longest and most variable commutes.

As with manufacturing, jobs in wholesaling, public utilities, and transportation are highly concentrated, extending from SoDo (South of Downtown), the site of most Port of Seattle operations, down the Duwamish and the Green River Valley, with the greatest concentrations in the city of Kent, and then around westward to the Port of Tacoma, equal in activity to the Port of Seattle. But the largest concentration of workers is in and around Sea-Tac airport. The residences of these workers are overwhelmingly in the same general areas, from south Seattle through all of south King County into Pierce County. Significant numbers of workers live in the city of Federal Way, itself with few jobs but a modest commute to Sea-Tac.

The location of jobs in education dramatically illuminates the significance of the University of Washington (the state's third-largest employer). Otherwise, education employment broadly reflects the distribution of the population and the schools and colleges that serve it. The residential location of health workers is broadly similar. We do not

2.9 Link Light Rail in downtown Transit Tunnel.

know the location of residence of government workers, but the location of jobs reveals four kinds of settings: military reservations, as at Fort Lewis–McChord (Pierce County), the Puget Sound Naval Shipyard (Bremerton) and Bangor submarine base (both in Kitsap County), and the Naval Station Everett; large prisons and institutions for the disabled and mentally ill; the Native American Indian Reservation operations; and general government offices, most dominantly in downtown Seattle (federal, state, King County, city of Seattle).

Commuting

Commuting still accounts for about half of daily traffic and, since it is concentrated in morning and evening rush hours, is responsible for most congestion. The mean commute

TABLE 2.2 COMMUTING MODE AND TIME, 2000-2006

CENTRAL PUGET SOUND	2000				2006			
Area	% SOV	% carpool	% transit	Avg Time	% SOV	% carpool	% transit	Avg Time
King	69	12	10	26.5	66.5	10.1	10	26.7
Pierce	76	13	2.7	28.4	78	10.6	4.4	27.9
Snohomish	75	14	4	29.6	75	14	4.1	30.6
Seattle	56.5	11.2	17.6	24.7	55	9.7	18	25.2
Bellevue	74	11	7	21.6	70	13.6	8.7	21.6
Everett	69	17	4	27.2	75	14.4	4.7	25.5
Federal Way	75	15	6	30.2	67	15.7	7.7	28.6
Kent	74	15	6	28.7	73	16	5	28
Tacoma	72	14	5.3	25.4	76	12.1	4.3	24

SOURCE: Puget Sound Regional Council, American Community Survey.

2.10 Mean commuting time in minutes, 2000.
Source: U.S. Census.

2.11 Shares of Single Occupancy Vehicles in census tracts, 2000.
Source: U.S. Census.

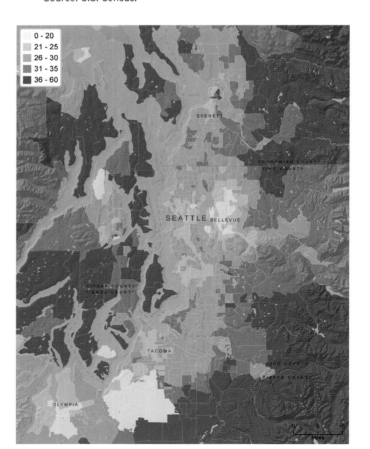

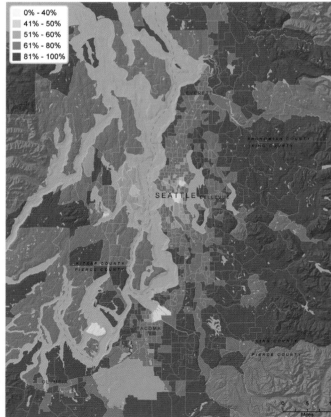

time for the geographically complex Seattle region is a not very high or unusual twenty-eight minutes (see table 2.2; fig. 2.10 shows mean commuting time between home and work from the 2000 census). Shorter commute times prevail in central and north-central Seattle, dominated by professionals, students, and service workers, in much of the cities of Tacoma and Everett, and especially and perhaps surprisingly, in the Eastside core cities of Kirkland, Redmond, and Bellevue. Conversely, longer commutes, over a half hour, characterize most if not all far suburban and rural areas, especially areas like Covington, Maple Valley, Tacoma's South Hill, Federal Way, and Marysville, and even southwest and southeast Seattle, which offer more affordable housing but insufficient jobs for the less affluent.

It is no surprise that single-occupancy vehicles (SOV) have such a high share of commuting modes (see fig. 2.11).

2.12 Shares of public transit users per census tract, 2000. Source: U.S. Census.

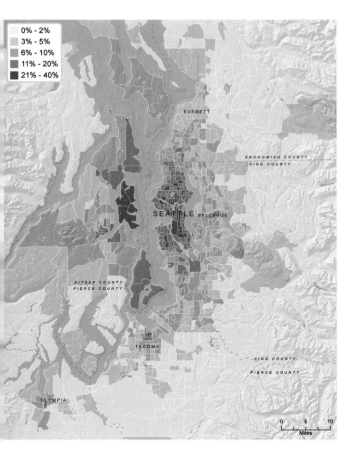

The highest shares occur in lower density outer suburbs, less well served by public transit, while the lowest levels (but still rather high) occur within the city of Seattle, with its historically strong transit system. Shares for carpooling are highest—from 20 to 40 percent—in the areas with the poorest working class populations, in general along the SR-99 corridor, in Kent and Auburn, in South Tacoma, and from southeast Seattle to Renton. These are typically households with only one vehicle. Carpooling shares are very low in areas with the highest use of public transit, essentially the area around Lake Union, from downtown Seattle to the University of Washington. Shares are also low in rural areas, with high shares of single-occupancy driving.

Highest shares of persons using public transit virtually coincide with the city limits of Seattle, except for very high levels in a few suburban locations, most notably Bainbridge Island, with high shares of passengers commuting by ferry (see fig. 2.25). The high shares in Seattle also reflect the mutual concentrations of professional jobs and educated residents, and such programs as transit subsidies for the elderly and for students and staff of the University of Washington (U-Pass).

LOCAL ENTREPRENEURSHIP

James W. Harrington and Charles Kauffman

Why has Seattle been such a successful place for entrepreneurship? To answer the question, we combine economic geography's oldest foci—physical geography and natural resources, and access to resources and markets—with contemporary economic geography's emphasis on innovation. We look specifically at the emergence of some of the Seattle region's best-known companies: Alaska Airlines, Amazon, Boeing, Costco, Microsoft, Nordstrom, Starbucks, and Weyerhaeuser. Along with the University of Washington, these world-recognized organizations are the area's largest employers (table 2.3), reflecting their brands' strengths and consumer presence. (Boeing, of course, does not sell to consumers, but it manages to enter the popular con-

TABLE 2.3 LARGEST EMPLOYERS IN WASHINGTON STATE, 2008

(**Bold** = headquartered in the Seattle-Everett-Tacoma-Bremerton CMSA)

Entity	Washington State Employment	Year Established in Washington
The Boeing Company	74,100	1916
US Army Fort Lewis	**40,091**	**1917**
Microsoft Corporation	**36,405**	**1979**
Navy Region Northwest	**23,961**	**1891**
University of Washington	**20,605**	**1861**
Wal-Mart Stores, Inc.	17,389	1993
Providence Health & Services Washington	**14,000**	**1856**
Fred Meyer Stores	12,788	1972
King County Government	**12,586**	**1852**
City of Seattle	**9,946**	**1869**
Group Health Cooperative	**9,185**	**1947**
MultiCare Health System	**8,552**	**1882**
Costco Wholesale Corp.	**7,475**	**1983**
Weyerhaeuser Corporation	**6,770**	**1900**
Alaska Air Group, Inc.	**6,565**	**1951**
Washington Mutual Inc. (no longer independent)	6,200	1889
Washington State University	5,725	1890
Starbucks Corporation	**4,884**	**1971**
Amazon.com[a]	4,800[a]	1984
Safeway, Inc.	4,673	1923
Lowe's Companies, Inc.	4,600	1990
Nordstrom, Inc.	**4,421**	**1901**
Swedish Medical Center	**3,860**	**1910**
Fairchild Air Force Base	3,723	1942
Qwest	3,639	1890s
Battelle-Pacific Northwest National Laboratory	3,388	1965

NOTE: Employment in the four-county Seattle metropolitan area totaled 1.8 million in 2007, according to the Puget Sound Regional Council (Puget Sound Trends, August 2008).

[a]Amazon.com does not disclose employment totals by location or region. A news article announcing the company's planned headquarters move noted that "city planners estimated this fall that Amazon could bring 6,000 employees to South Lake Union over the next five years" (E. Pyrne, "Amazon to Make Giant Move," *Seattle Times*, 12/22/97; http://seattletimes.nwsource.com/html/businesstechnology/2004087320_amazon22.html). This table reduces that estimate by 20 percent, to allow for the estimate's assumption of employment growth.

sciousness by way of its duopolistic position in producing commercial jetliners). These companies trace their origins to twentieth-century Seattle. They have retained their corporate independence and continue to grow. Can we identify anything about Seattle and the Puget Sound region that accounts for the origins and rise of these companies?

Key Processes: Physical Geography and Innovation

The salient aspects of the region's physical geography are its mild climate, abundant forest resources, and deepwater harbors. The salient aspects of its accessibility are those harbors, its distance from other major metropolitan centers, and its status as the largest U.S. metropolitan area closest to Alaska (fig. 1.9). As noted in the previous section, in the nineteenth century these attributes resulted in the area's selection as a major railroad terminus, fortunes made from exploiting the forests, and the attraction of people and wealth from the Klondike Gold Rush (1897–98). Logging and lumber were the basis for a number of the region's other large employers, including Weyerhaeuser. The founding of Boeing and Nordstrom results indirectly from these features, with the link being entrepreneurs whose capital came from lumber (William Boeing) and the Yukon (John Nordstrom). But there is more to it than just these traditional aspects of economic geography.

Successful innovations may be expensive and rare, but the market pays a premium to the one or two firms that are first-movers. These super-profits can be large enough to change the economic face of a region, as steel did in Pittsburgh, semiconductors did in Palo Alto, and airplanes and software did in Seattle.

In studying long-term regional economic change, we emphasize "innovative entrepreneurship," defined as developing an economic activity that is new to the economic system (at regional, national, or global scale, depending on the purpose of the analysis). The new activity can be manifested within organizations or especially through the founding of new firms; in either case, it represents the introduction of a new sector or activity into the system. Thus, we do not emphasize the more common motivations for new-firm foundation, that is, many small firms or sole proprietorships are begun only because the entrepreneur cannot find or does not want wage or salary employment but has access to sufficient capital, expertise, and time to start an independent operation.

Studies of entrepreneurship suggest six possible bases for the geographic concentration of new ventures over time:

◊ manager or employee spinout from existing establishments, which usually happens in the same region;
◊ social or intergenerational familiarity with entrepreneurship;
◊ localized social norms leading to acceptance of and support for self-employment, growth of small organizations, and even failure;
◊ early awareness of new technologies and their possible applications, based on entrepreneurs' exposure to innovations and to potential markets;
◊ early-stage capital sourced through (often localized) networks; and
◊ intergenerational transfers of wealth.

Case Histories

WEYERHAEUSER

Emigrating from Germany in 1852, Frederick Weyerhaeuser soon discovered a talent for entrepreneurship and the economic potential of North American forests. After a few years on the East Coast, he found work at an Illinois sawmill. Hired as a night watchman, Weyerhaeuser used one of the industry's downturns as an opportunity to purchase the mill. Following the trail of timber to its source eventually led him to the Pacific Northwest, where in 1900 he acquired 900,000 acres of timberland from the Northern Pacific Railway and founded the Weyerhaeuser Company. Washington's abundance of timber resources and the availability of a diverse network of transportation methods made the region a perfect center for the timber industry.

Following Frederick's death in 1914, his son F. E. Weyerhaeuser took control of the company. F. E and his son J. P. were able to make Weyerhaeuser Company responsive to changing market conditions. The 1930s marked the beginning of tree-plantings to replace natural reforestation, and the advent of the Pres-to-Logs to utilize scrap wood. The construction of the world's first pulp mill in 1931 kept Weyerhaeuser afloat during the Great Depression through the production of paper, while most operations sustained major losses.

Acquisitions such as MacMillan Bloedel of Vancouver, B.C., and Willamette Industries of Portland, Oregon, have given Weyerhaeuser a strong presence throughout the region, along with a diverse offering of products and services. The proximity of deepwater ports enabled trade with developing markets in Southeast Asia and South America, allowing Weyerhaeuser to become the dominant world provider of lumber.

2.13 White Pass & Yukon Route, 1905.

NORDSTROM

Emigrating from Sweden in 1886, John W. Nordstrom arrived in New York with five dollars. A series of labor jobs carried him westward to Seattle, where he first learned about the Klondike gold fields in Alaska. He used his hard-earned savings to buy a one-way ticket to Skagway, where he staked claim to a productive field that he was able to sell for $30,000. In 1900 he returned to Seattle, which had a milder climate than Alaska and yet good water access to the gold fields (fig. 2.13). While looking for a way to invest his fortune, John encountered a friend he had made in Alaska, Carl Wallin. Wallin and Nordstrom decided to partner and together they opened a small shoe-repair shop. The company's expansion was slow: Nordstrom's second shop was opened two decades later, in 1923.

John Nordstrom retired in 1928. His two children bought out Wallin's stake in the company, and intergenerational transfers of ownership have kept the company in the family and brought new approaches to the business. New

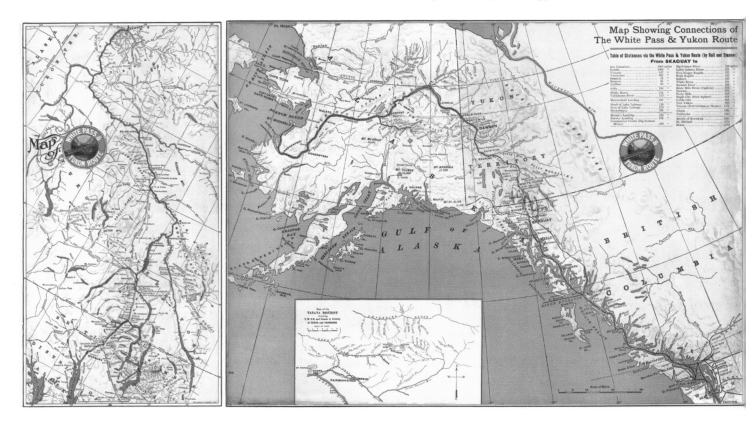

2.14 Senator Henry M. Jackson in front of the Boeing Red Barn, 1982.

product lines were added through the acquisition of clothing retailers, and Nordstrom became a department store. With the third generation leading the company, Nordstrom went public in 1971 and subsequently began its expansion across the United States.[20] In 1985 Nordstrom became the largest specialty store chain in the country and now operates over 150 stores in twenty-eight states.[21]

BOEING

The son of a German immigrant who gained wealth from trading land, timber, and iron ore, William Boeing benefited from a childhood of prosperity. He studied engineering at Yale, leaving at age twenty-one to seek his own fortune in the burgeoning timber industry of Washington State. He began two timber companies and moved from the Pacific Coast to Seattle, where he bought a shipyard in 1910.[22] He developed an interest in flying machines and traveled widely to view them. In 1915 he purchased and assembled an airplane, grew dissatisfied with the way it flew, and

worked with naval engineer Conrad Westervelt to design a better plane. The next year the two men started Pacific Aero Products Company, which soon became Boeing Airplane Company. Figure 2.29 shows Boeing's original manufacturing facility. William Boeing relied on his boat craftsmen to build the wood-framed planes, utilizing the wood-working expertise in this lumber-rich region.

Boeing's commercial aircraft design and production have been focused in the Seattle region for all of the company's history, while its military production has been dispersed—in part reflecting the locations of its corporate acquisitions, and in part reflecting the preference of the U.S. government to produce some military aircraft away from the West Coast. However, the company has threatened to relocate production of new airplane models throughout its history, thereby receiving major public infrastructure and tax benefits from Washington State and local municipalities. Three recent actions have increased the company's bargaining power: the relocation of corporate headquarters to Chicago in 2001; its unprecedented reliance on external contractors to design, build, and finance major components of the two-year-delayed 787; and its 2009 purchase of fuselage-manufacturer's facilities in South Carolina to allow its first commercial-aircraft assembly outside its home region. Boeing has made clear its strategy to become a global engineering corporation rather than a Seattle airplane manufacturer, with implications for its local linkages and potential for local entrepreneurship.

ALASKA AIRLINES

With its headquarters in Seattle and market focus in California and Mexico, Alaska Airlines evokes a sense of geographic irony. Mac McGee purchased his first plane in 1932 from a Boeing subsidiary and formed McGee Airways to provide air transport services in Alaska. The company received a contract to deliver air mail, but Anchorage's harsh winter climate prevented the delivery of large packages during the winter. The airline established a storage facility in Seattle to house the packages until they could be delivered; again, Seattle's climate and relative location proved favorable. After mergers, the company was renamed Alaska Airlines in 1944. The Civil Aeronautics Authority limited the airline's market area primarily to Alaska until airline deregulation in 1978. Then the Alaska Air Group was founded in Seattle, where its headquarters and operations hub remain.

STARBUCKS

The company known for keeping the world awake and alert, Starbucks also got its start in Seattle. In 1971 a team of sleep-deprived professors opened the first Starbucks inside Pike Place Market, the oldest continuous public market in the United States. The company roasted and sold coffee beans for home preparation. In 1981 a salesman for a Swedish housewares manufacturer called on the Seattle coffee shop. After trying the dark roasted coffee, Howard Schultz fell in love with Starbucks and convinced the owners to take him on as a business partner.

Visiting Italy in 1982, Schultz was struck by the small espresso bars on nearly every street corner, offering freshly prepared drinks and a quaint atmosphere in which to relax and unwind. Could Starbucks coffee be prepared on premises, in many locations? His Seattle partners refused to change the company's business model, so Schultz created a new company, Il Gironale, modeled on the Italian espresso bar. The following year, Il Gironale purchased Starbucks's four existing stores and was rebranded as Starbucks. With its headquarters remaining in Seattle, the company underwent immense expansion—for several years opening a new store nearly every weekday. Today the company operates 16,120 stores in fifty countries.

MICROSOFT

William (Bill) Henry Gates III was born into a legacy of entrepreneurship, law, and education. William Henry Gates Jr., Bill's father, had earned a bachelor's degree from the University of Washington in 1949 and a law degree the year after. In 1964 he joined the law firm Shidler and King (later Preston, Gates, and Ellis). The firm's success made it possible for the senior Gates to send his children to Lakeside School, one of the few places where young Bill could have become acquainted with basic computers amid a rich network of relational capital.[23] Using University of Washington computers, Bill and his classmate, Paul Allen, developed a program to analyze traffic data. As word of the traffic application spread, the Bonneville Power Administration requested Bill and Paul's assistance to computerize the Northwest power grid.

After graduating from Lakeside, Paul Allen went to Washington State University in Pullman, while Bill enrolled at Harvard in 1973. Paul Allen dropped out of Washington State as a sophomore to take a programming job with Honeywell in Boston. He convinced Gates to leave Harvard and help him program a language for the new, relatively affordable Altair 8800 computer. Together, they founded Micro-soft in 1975, near Altair's manufacturing plant in Albuquerque, New Mexico. In 1978 Microsoft, minus the hyphen, left Albuquerque for Bellevue, Washington.

The small company benefited from another company's new organizational model and new product type. Giant IBM was beginning development of a stand-alone computer for individual workstations—the "personal computer" (PC). IBM decided to change its business model of internal production of proprietary computers and operating systems to develop the operating system for a new type of machine that any manufacturer could produce—and IBM selected Microsoft. By 1982 Microsoft had licensed its M-DOS operating system to over fifty personal computer manufacturers. After many generations of new operating systems and end-user software, Microsoft software is presently on 90 percent of PCs worldwide.

COSTCO

Transfers of entrepreneurial skills have played a key role in the founding of the leading wholesaler in the United States. Jeff Brotman was the son of Bernie Brotman, who created eighteen stores throughout the Northwest. Jeff Brotman began his career by partnering with his brother to form a specialty jeans retailer in the late 1970s. He continued investing in and opening retail establishments in the Northwest, but it was not until 1982 that he and James Sinegal opened the first Costco. The Costco concept, a no-frills membership club reliant on customer loyalty, was the brainchild of Bernie Brotman. While Costco began by selling products to small businesses, new levels of club membership made the company popular among families. Costco operates its recession-resilient business model through 520 stores and online.[24]

AMAZON.COM

And, finally, we present a company that was founded in Seattle, but was based on ideas and wealth from an entirely different sector in an entirely different place. Jeff Bezos came from a family of scientists. He earned degrees in electrical engineering and computer science from Princeton in 1986. While working as a senior vice president for Bankers Trust in New York, Bezos developed a highly sophisticated hedge fund through which he discovered the economic potential of the Internet. His entrepreneurial breakthrough was to combine his awareness of computer science, the

Internet, and finance. In 1993 he and his wife left New York and drove across the country to Seattle, where they founded Amazon.com a year later, drawn to Seattle's proximity to two major book wholesalers and the city's pool of software talent.[25] The company's early-adopter status and aggressive growth provided a trade name, customer base, software, data, and systems infrastructure to allow it to survive the "Internet bust" of the early twenty-first century.

Geographies of Entrepreneurship?

Though the above brief review omits several large organizations that form parts of the region's economic base, it illustrates the variety of influences on innovative entrepreneurship. Weyerhaeuser's original success required huge timber resources; its continued success resulted from corporate acquisitions and diversification. Nordstrom's founding reflects development derived from the natural resource of gold and the access that Seattle provided to Alaska and the Yukon. Nordstrom's original capital came from gold mining; however, the remote location of the gold fields encouraged the related final-demand sector (shoes and gear) to be located elsewhere. The chain's subsequent growth was based on entrepreneurship fostered through intergenerational wealth.

Intergenerational wealth had a direct role in Bill Boeing's ability to become enthralled with flying and his ability to begin an aircraft company; that wealth was based in his and his father's exploitation of resources (in the Northwest and the Midwest, respectively). However, the Boeing Company grew from the entrepreneur's early awareness of a new technology and his development of new markets—military and air delivery. Despite nearly a century of local linkages (specialized and experienced labor; a network of relatively small supporting firms; public support of education, training, and infrastructure), it is arguable whether an innovative milieu has been established external to the company, and the company has redoubled its efforts to maintain locational flexibility. Alaska Airlines also got its start in Seattle because of demand in and from Alaska that was more easily satisfied from operations based in Seattle's milder climate. Microsoft's founding reflects the entrepreneurs' early exposure to a new technology and ability to apply it to develop a new sector—software sales to computer manufacturers and end users. There was synergy with an external firm's decision to start a new subsector, personal computers.

Localized, intergenerational entrepreneurship—the ability of one generation of a family to learn about entrepreneurship and about a type of economic activity from the previous generation—was one element in the development of Costco. However, successful entrepreneurs raise children in every region; not every daughter or son develops a new organization with a new business model. The Costco case bears further investigation.

Technological exposure and awareness of financing possibilities underlay Jeff Bezos's founding of Amazon and Howard Schultz's purchase of Starbucks. Retailing and coffee making were centuries-old activities, but the application of new technologies and creative financing were the bases for entrepreneurial success. The Seattle economy was the lucky beneficiary of these flexible location decisions.

These examples of innovative entrepreneurship, bringing new economic activities to the region, resulted from physical geography in the early twentieth century, relative location and natural resource–based wealth in the first third of the twentieth century, and individuals' early exposure to new technologies in the last third of the century. What regional characteristics will lead to innovative entrepreneurship in the current century?

FROM WHEAT TO WINE:
INSTITUTIONAL CHANGE AND REGIONAL DEVELOPMENT IN WALLA WALLA

Nicholas Velluzzi

....

James Harrington and Charles Kauffman discussed the unique combination of factors in the Seattle area that nur-

tured entrepreneurship, but my own discussion focuses on a very different part of the Northwest: the Walla Walla Valley in Eastern Washington (fig. 2.15). And while this east of the mountains place may seem far away, it has a significant effect on Seattle—and vice versa—as illustrated in figure 2.16. Over the past thirty years or so, the Walla Walla regional economy has been undergoing a steady transformation, from that of a lagging, peripheral, largely wheat-farming region to an economically dynamic one. Nowadays mentioning Walla Walla will suggest *wine* to the listener. From 1977 until 1991, there were only six wineries operating in the region. By 2000 the number had increased to twenty-three. By 2010 there were well over a hundred commercial wineries in the Valley. A similar pattern applies to vineyard expansion. When the Walla Walla Valley appellation was certified in 1984, there were approximately 80 acres dedicated to wine grapes; in 2006 the planted vineyard acreage amounted to more than 1,500, with an additional 1,000 acres prepared for planting in 2007.

Given the Walla Walla Valley's relative isolation, how can we understand the emergence and eventual growth of a major wine industry there? What were the social processes at work that facilitated the shift in the valley's develop-

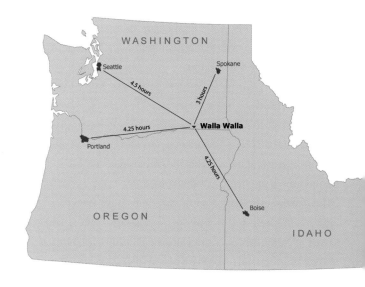

2.15 Location of Walla Walla with reference to the Pacific Northwest.

2.16 September 2009 cover of *Seattle Metropolitan* magazine. Though Walla Walla is 270 miles from Seattle, its changing economy is having cultural and economic effects throughout the region. Reprinted with permission.

ment path from wheat to wine? What enabled Walla Walla to "catch up" to other Old and New World centers of wine production and acquire a reputation for "world class" wine production?

Establishing a new economic development pathway entailed a break from the old ways of commodity production and the creation of new institutions and frameworks governing production practices, product quality, industrial performance, and competition. Even though there are tremendous differences among wine, wheat, and canned fruit, they share common roots in farming and manufacturing. It is no coincidence that Walla Walla's pathbreaking winemakers were a tight-knit community of individuals whose livelihoods were based in farming and food processing. From a local and regional development standpoint, it is interesting to take note of the way in which that group

combined and mobilized its knowledge resources and experience to create a new, distinct set of rules or structures to shape production practices for winemaking. Below I provide a set of examples of industry-specific organizations that were created in the valley for the purpose of building and articulating the institutions and conventions that constitute the Walla Walla wine industry.

Economic change in Walla Walla was driven by a combination of individual strategic action and institutional innovation, both of which resulted in the production of new institutional arrangements designed to facilitate the economic development of the local wine industry. There are four examples that illustrate this process.

First, the Walla Walla Valley was certified as an American Viticultural Area in 1984. AVA is a federal classification that governs place-of-origin labeling and product identity (i.e., AVA designation regulates the conditions under which vintners can label their wine "Walla Walla Valley"). It is important to think of place-based product identity as being shaped by three things: local commodity production, consumption in distant markets like Seattle, and a reputation of quality associated with the region and its producers. The majority of Walla Walla wineries rely on markets in the Northwest's urban centers, though some producers have established markets across the United States, Europe, Japan, and China.

Second, the Walla Walla Community College Center for Enology and Viticulture was established in 2000 for the purpose of providing industry-specific training and degree programs. One of the first programs of its kind, the center hosts an on-site teaching winery that has been emulated by similar programs throughout the Northwest and beyond.

Third, the Walla Walla Valley Wine Alliance is an industry association for locally based wineries, grape growers, and related producers, such as suppliers. Established in 2001, the Alliance provides a means for producers to address industry-specific problems, helps market the appellation, and enables the industry to speak publicly with one voice.

2.17 A Walla Walla winery in summertime.

Last, in 2005 the Walla Walla Valley VINEA (a group of winegrowers who have embraced environmental and social sustainability for their wineries) was created to promote what they call a sustainable trust. By attempting to shape the approach to viticulture within the appellation, VINEA is simultaneously creating conventions of quality that are strategically oriented toward a particular kind of consumer who seeks products identified as "*terroir*-driven," that is, sustainable, natural, and noninterventionist. All of those institutional arrangements emerged in tandem with the growth and maturation of the industry. By illustrating the interplay between individual strategic action and institutional innovation over time, the Walla Walla wine industry created institutions from place-specific conditions and structures that shape how actors, in this case wine producers, engage in commercial winemaking. It is important to note that although those institutions are born locally, they are shaped by and interact with other institutions that operate at and across a variety of spatial scales, such as in the legal environment in Washington State and the United States, through conventions of quality in other wine-producing regions and according to consumer tastes in local and distant markets. Though Walla Walla is the locus of my research, my aim is to broaden the apprecia-

tion and treatment of individual agency, institutions, and economic change in order to improve our understanding of widespread and uneven geographical development.

TRADE WITH CHINA

Kam Wing Chan and Spencer Cohen

Trade between Seattle and China has been long-standing and formative to the region. In 1868, a year before Seattle's incorporation, Chin Chun Hock founded the famous Wa Chong Company, a U.S.–China trading house, which continued into the twentieth century.[26] In the last thirty years, every Chinese leader, including Deng Xiaoping, has made a point of spending some time in Seattle, and trade has always been an important agenda item on their visits. Several important administrative and diplomatic partnerships have also developed in Seattle since the reopening of U.S.–China economic relations more than thirty years ago. The Washington State China Relations Council, based in Seattle, was founded in 1979. The City of Seattle has since 1983 maintained a sister city relationship with the municipality of Chongqing, and three of the Port of Seattle's sister-port relationships are with Chinese ports: Shanghai (1979), Qingdao (1995), and Dalian

2.18 Port of Seattle.

(2007). In addition, the U.S. office for the U.S.–China Clean Energy Forum, an initiative to promote joint problem-solving and collaboration on developing clean energy solutions, is also housed in Seattle.

In 2008, total two-way trade (imports and exports combined) between the Port of Seattle and the People's Republic of China (excluding Taiwan, Hong Kong, and Macao) surpassed $19.2 billion, constituting almost half (48 percent) of all two-way trade through the port (fig. 2.19). This amount includes imports and exports to and from other parts of the United States,[27] reflecting the general national picture of much greater imports than exports. Seattle has arguably been one of the largest beneficiaries of China's economic boom over the last two decades. With its close geographic proximity to Chinese ports relative to other West Coast regions, Seattle has prospered in recent years as a major hub for U.S.–China international trade flows. The port authorities of Seattle and Tacoma together constitute one of the busiest container loading centers in the United States, largely owing to Washington's advantageous position in the Northwest and proximity to China's manufacturing centers and rail linkages with the American Midwest. Seattle was the port-of-call for the first Chinese cargo ship to enter the United States in 1979.

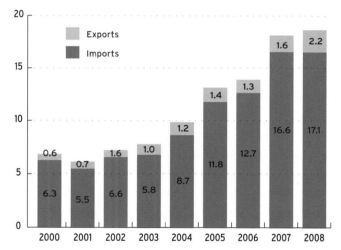

2.19 Two-way China trade through the Port of Seattle, 2000–2008. Source: Port of Seattle, Foreign Waterborne Trade Report.

Washington State–China trade flows effectively "took off" in the last decade. Washington's state-of-origin goods passing through the seaports of Seattle and Tacoma increased nearly twenty-one times in the period between 1997 and 2008. Though there is no sound methodology for estimating a state-level current account, some have estimated that Washington is the only state in the Union that runs a trade surplus with China, in contrast to the national pattern of huge U.S.–China trade deficit.[28] Measuring "trade" accurately at the subnational level is a difficult, if not impossible, task, as few produced commodities these days are made in only one single location. However, counting trade based on the origin of assembly and excluding "pass-throughs,"[29] the value of goods exports (as opposed to services exports) originating in Washington State, including many expensive aerospace and computer products, is very likely higher than that of products shipped from China and consumed in the state, even in large quantity.

In 2008 Washington State's exports to China reached $5.8 billion,[30] making China the state's third largest foreign market that year. On a per capita basis, $891 worth of goods was sold to China per each Washingtonian, up from $304 in 2000, placing Washington second only to Alaska by this measure.[31] Washington was the largest individual state exporter to China in aerospace products ($4.2 billion), fur skins and artificial furs, and prepared fruits and nuts, and ranked second among all fifty states in exports to China of ores, slag, and ash; seafood; and wood products.

Aerospace goods, primarily Boeing airplanes, have made up the lion's share of Washington's exports to China, averaging 81.5 percent per year between 1997 and 2008; over this period, aerospace exports grew roughly 168 percent.[32] It

THE FIFTY-YEAR CAREER OF AN ECONOMIC GEOGRAPHER

William Beyers

Upon my retirement from the University of Washington at the end of the spring quarter of 2010, the editors of this volume asked me to write about my research in economic geography, which I will describe briefly here. I have focused on the geography of the Seattle region, in both basic and applied research, and on the Washington economy since the mid-1960s and published in a wide variety of academic and public policy media. The research has often straddled both basic and applied approaches, with basic research informing policy direction and with the needs of policymakers being published for academic readers.

I have been involved with the measurement of many of the input-output models developed for Washington State. These models track the markets and sources of supply of all industries in the state economy. This unique series of seven models—benchmarked in 1963, 1967, 1972, 1982, 1987, 1997, and 2002—provides an unparalleled overview of the changing structure of the Washington economy. The models clearly show the growing importance of foreign markets for Washington industries, the growth and diversification of service industries, and the growing importance of interregional trade in services. They have been used in dozens of economic impact studies across the region—on subjects ranging from national parks in Washington State to Mariners baseball, the high-technology industry, arts and cultural organizations, and business trade shows.

I have taken part in economic forecasting and planning in the Northwest, including consulting about electrical power demand in the city (for Seattle City Light) and in the Pacific Northwest (for the Northwest Power and Conservation Council).[35] I participated in the research that led to the city of Seattle abandoning its involvement with Washington Public Power Supply System (WPPSS) nuclear plants numbers 4 and 5. That decision reformed the electrical power planning framework in the entire region.

Richard Morrill (coeditor of this volume) and I have worked on two projects that have influenced regional planning. We developed measures of the unmet need for higher education enrollment in Washington State, which resulted in the recommendation for the location of two branch campuses developed in the last two decades for the University of Washington. We were also on the Growth Strategies Commission, whose recommendations were used by the Washington State legislature to create the 1990 Washington State Growth Management Act.

In the mid-1980s, my subject of research was producer-services businesses that sold their services primarily to is no coincidence that when China's President Hu Jintao visited Seattle in 2006, the luncheon held in his honor (which both authors attended), took place in Boeing's Aviation Center Gallery. Former President Jiang Zemin also paid a call to a Boeing worker's home when he was in town attending the APEC meeting in 1993. China has been a major purchaser of Boeing aircraft whose final production takes place here: the 737, 747, 767, and 777, as well as the new 787.

However, *nonaerospace* manufacturing has also been growing rapidly as a component of Washington's exports to China, illustrating the increasing diversity of trade relations between the state and China. Between 1997 and 2008,

the monetary export value of nonaerospace manufacturing goods grew an astounding 411 percent, at an average yearly pace of roughly 16 percent. The largest components of this growth have been in computer and electronic products, food processing, and nonelectrical machinery.

There have been several factors driving these trade flows. As a West Coast city with a large Asian population, Seattle has traditionally looked to East Asia for business opportunities and new markets. A large segment of Washington's economy is also naturally export-oriented. Aside from Boeing planes, the largest exports in the United States are wheat, fruits, and processed foods (e.g., French

businesses and government, rather than to households. One of my research assistants on that project, Mike Alvine, worked for the Central Puget Sound Economic Development District, where he recognized that producer-services were growing rapidly at the same time that a key local industry (aerospace) was in a downturn. Conventional wisdom at the time suggested that local services were linked to companies like Boeing and should have been drawn down in the aerospace business cycle. So we surveyed more than a thousand local producer-services businesses and found that they had strong markets outside the local area and had only a minor dependence on local manufacturers.

These surprising findings led to a long-standing interest in producer-services. My research showed (1) strong niche markets—firms strongly differentiating themselves from their competitors; (2) divergent bases for competitive advantage—firms much more frequently emphasizing their quality as opposed to the low cost of their services; (3) firms with increasing levels of export markets; and (4) strong personal contact structures in the process of producing and delivering services. This research did not find a decrease in face-to-face contacts as new digital modes of contact became possible, but what we did find was a strong entrepreneurial drive in the process of starting firms and an increasing use of part-time and temporary help in the production process.

I have always enjoyed experimenting with research methodologies. In a recent collaboration with Professor Paul Sommers of Seattle University, we developed a different set of industry "clusters" for the Central Puget Sound region than the ones that were being used by consultants for the Puget Sound Regional Council (PSRC). We were critical of the narrowness of the PSRC cluster definition, as it covered only about one-third of employment in the regional economy, while our analyses covered the entire economy.

I have also enjoyed tackling research aimed at helping with public policy issues and environmental causes. I helped evaluate (with others) fiscal impacts of the Growth Management Act and the likely level of claims against the State of Washington associated with Initiative-933, a measure that would have required compensation for "takings" associated with aspects of the Growth Management Act. I have been actively involved with the Alpine Lakes Protection Society (ALPS) since the 1970s, and was president of the association when Congress declared the Alpine Lakes a wilderness area. And, like any good geographer, I drew on my cartographic skills to help ALPS publish a map of the Alpine Lakes Wilderness, which has produced a steady stream of revenue to this small nonprofit environmental organization.

fries), for which Washington is both major producer and employer within the Greater Seattle area. Beginning in the early 1980s, Greater Seattle has seen the emergence of a major cluster of information, communication, and technology (ICT) firms anchored by Microsoft and later Amazon and RealNetworks, and robust telecom and life sciences industries. With a state population of only 6 million, these firms have always looked outside the region for business development and expansion opportunities. Historical and cultural linkages with China have further enabled these business ties to grow.

Note that these data mask the important and significant role of *services exports* to China (e.g., education, legal services, overseas architecture contracts, etc.). For example, in 2008, the share of international students from China attending schools in Washington State increased from 8 percent to 10 percent. Many of these schools, both community and technical colleges and universities, reside in the Greater Seattle region, including Seattle proper, Bellevue, and Everett. Several world-class Greater Seattle architecture firms have done significant design work in China (e.g., Mulvanny G-2), and a growing number of Seattle-based law firms have established China operations (including Davis Wright Tremaine, where former Washington governor Gary Locke was a partner before joining the Obama Administration's cabinet in 2009).[33]

In recent years, foreign direct investment has emerged as another key aspect of Seattle-China relations. Several Chinese ICT companies have established overseas operations in the Greater Seattle region, including iSoftstone and ChinaSoft. Mindray International, a Shenzhen-based medical device company, chose this region for its first foray into the U.S. market, establishing its North American headquarters and research and development facility in Redmond, Washington (headquarters were later moved to the East Coast following the firm's acquisition of a New Jersey company). In October 2009, the Chinese company Modern Dental Laboratories established a headquarters for U.S. operations in Bellevue and a service center in Seattle under its new subsidiary, Modern Dental Laboratories USA.[34]

NOTES

1   Charles Tiebout, *The Community Economic Base Study* (New York: Committee on Economic Development, 1962); D. North, "Location Theory and Regional Economic Growth," *Journal of Political Economy* 63 (1955): 243–58; J. E. Vance Jr., *The Merchant's World: The Geography of Wholesaling* (Englewood Cliffs, NJ: Prentice Hall, 1970).

2   S. Mickelson, *The Northern Pacific Railroad and the Selling of the West* (Sioux Falls, SD: Center for Western Studies, 1993).

3   D. H. Clark, *An Analysis of Forest Utilization as a Factor in Colonizing the Pacific Northwest and In Subsequent Population Transitions.* PhD diss., University of Washington, 1952.

4   Boeing Company, *The Boeing Logbook, 1916–1991* (Seattle: Boeing Historical Archives, 1992).

5   R. S. Kirkendall, "The Boeing Company and the Military-Metropolitan-Industrial Complex, 1945–1953," *Pacific Northwest Quarterly* 85 (1994): 137–49.

6   D. McClary, "Puget Sound Naval Shipyard," 2003, http://www.historylink.org/index.cfm?DisplayPage=pf_output.cfm&file_id=5579 (accessed February 22, 2010).

7   H. Mansfield, *Vision* (New York, NY: Madison Publishing, 1986).

8   P. Blecha, "Conductor Sir Thomas Beecham Debuts with Seattle Symphony Orchestra on October 20, 1941," 2002, http://www.historylink.org/index=cfm?DisplayPage=pf_output.cfm&file_id=3877 (accessed February 22, 2010).

9   Vance, *The Merchant's World.*

10 William Beyers, "The Economic Impact of Technology-Based Industries in Washington State." Manuscript, Department of Geography, University of Washington, 2008.

11 Civil Aviation Forum, "Boeing 787 Production and Logistics Detailed," 2005, http://airliners.net/aviation-

forums/general_aviation/read.main/2320988 (accessed February 17, 2010).

12 Port of Seattle, "Seattle Harbor Ten-Year History of Cargo Volumes Handled: 2000–2009," 2010b, http://www.portseattle.org/seaport.statistics/pos10yearhistory.shtml (accessed February 17, 2010). Port of Tacoma, "Container Volumes," Container_YTD_2010-01[1].pdf.

13 William Beyers and David Lindahl, "Explaining the Demand for Producer Services: Is Cost-Driven Externalization the Major Factor?" *Papers in Regional Science* 75 (1996): 351–74.

14 P. Sommers, *Economic Impacts of the Military Bases in Washington* (Olympia, WA: Office of Financial Management, 2004).

15 Cindy Warner, "Seattle Opera Predicts 9.5 Million Dollar Economic Benefits from the 'Ring Cycle,'" *San Francisco Opera Examiner*, August 16, 2009.

16 William Beyers, Christopher Fowler, and Derek Andreoli, "The Economic Impact of Music in Seattle and King County." A report prepared for the Mayor's Office of Film and Music, City of Seattle.

17 Ibid.

18 CIC Research, Inc., *The 2008 Market Profile and Economic Impact of Seattle-King County Visitors* (San Diego, CA: CIC Research, Inc., 2009).

19 Port of Seattle, *Cruise Passengers*, 2010a, http://www.portseattle.org/seaport/statistics/cruispassengers.shtml (accessed February 17, 2010).

20 Seattle HistoryLink.org. Nordstrom Department Store. HistoryLink.org Essay 1677, http://www.historylink.org/index.cfm?DisplayPage=output.cfm&file_id=1677.

21 Nordstrom's 2008 Annual Report, https://materials.proxyvote.com/Approved/655664/20090311/AR_38144/HTML2/nordstrom-ar2008_0078.htm.

22 http://www.boeing.com/history/chronology/chron01.html.

23 R. Capello and A. Faggian, "Collective Learning and Relational Capital in Local Innovation Processes," *Regional Studies* 39 (2005): 75–87.

24 Retail Market Research Handbook, Warehouse Clubs," 2006, http://www.researchandmarkets.com/reportinfo.asp?cat_id=0&report_id=305390&q=warehouse%20clubs&p=1.

25 Academy of Achievement, "Jeff Bezos Biography." http://www.achievement.org/autodoc/page /bez-0bio-1, 2009.

26 Ron Chew, ed., *Reflections of Seattle's Chinese Americans* (Seattle: University of Washington Press and Wing Luke Asian Museum, 1994).

27 Port of Seattle, "2008 Waterborne Foreign Trade Report."

28 *The Christian Science Monitor*, "Five Cities That Will Rise in the New Economy," November 20, 2009, http://www.csmonitor.com/Money/2009/1120/five-cities-that-will-rise-in-the-new-economy (accessed November 24, 2009).

29 Those goods that were produced elsewhere and only pass through Washington and Seattle on their way to foreign markets.

30 For the remainder of this discussion, export figures will be corrected for "pass-throughs," or crops that actually come from elsewhere, such as soybeans, but are often listed as originating in Washington State because of consolidation before shipment. These include soybeans, corn, and tobacco products.

31 Calculations are based on U.S. Census Bureau population estimates for 2000 and 2008.

32 Data are from *Wiser Trade*, http://www.wisertrade.org.

33 See, for instance, Steve Wilhelm, "Asian Exchange: China Tie Lets MulvannyG2 Spread Its Design Wings," *Puget Sound Business Journal*, June 6, 2003.

34 Steve Wilhelm, "China Dental Lab Picks Bellevue to Bite into U.S. Market." *Puget Sound Business Journal*, September 25, 2009.

35 The Northwest Power and Conservation Council was created by the U.S. Congress in 1980; Beyers served on its forecasting committee in the early 1980s.

THREE

GLOBAL GEOGRAPHIES

Matthew Sparke

Globalization is a central focus of research in contemporary human geography. The increasing connections among more and more places on the planet and the accelerated pace at which interactions occur mean that understanding places must entail looking at their situation and representation in relation to global economic, political, and cultural ties. We saw hints of this approach in chapter 2, but in this chapter, Seattle's global geographies are the central focus, providing a way to see Seattle as a global city in relation to three different regimes of global interdependency: as a competitive city, a collaborative city, and a curative city.

Seattle is undoubtedly a global city, but of what sort? At the end of the 1980s, local business leaders promoted it as a globally competitive, livable city. Urban development and renewal were to be built on growing Pacific Rim trade, software and biotech innovation, and the remaking of the downtown as a spectacular world-class urban destination. A decade later, this same vision of the city as a "24/7" meeting place for trade partners and advocates, techies, and transnational tourists prompted a successful bid to host the 1999 World Trade Organization meeting. However, arriving along with the trade officials for the much-anticipated event were other global citizens with much more radical ideas about how to make the livable city world class.

Environmentalists, students, unionists, and a wide range of human rights groups organized to remake Seattle into an altogether different sort of global city. Seattle was still envisioned as a gateway for border-crossing transnationals, but in November 1999 the polyglot citizenry swarmed the streets to redefine the meaning of world-class livability in terms of global justice, democracy, and more collaborative approaches to global living. The resulting confrontation made Seattle renowned in the world as a contested global city, but, after the tear gas cleared, the visions of city and citizenship did not stop evolving. Thus today we see yet another global Seattle being built. Neither the promoters of market competition nor the collaborative proponents of global justice have gone away, but in the aftermath of their now-famous standoff, a third and arguably "curative" rethinking of the city is taking shape: a re-visioning of Seattle as a world center of global health philanthropy and other private-sector treatments for the mismatch between global markets and global justice.

From competitive global business city to collaborative global justice city to curative global philanthropy city: each vision of Seattle's global nature enables and exemplifies different forms of citizenship. These distinct forms of citizenship are important because they suggest how citizens are becoming *denationalized* in global cities.[1] Studying

a city such as Seattle provides a way of examining the so-called cosmopolitanism of citizenship on the ground. Following the argument of David Harvey, geographical research can also in this way reveal whose freedoms and what rights are at stake in metropolitan accounts of cosmopolitanism.[2] But, beyond academic debates over denationalization, the changing articulations of city and citizenship are also of immediate local importance because of the ways in which each incarnation of a "global Seattle" has been lived and thereby layered onto the city's landscape.

To understand the different articulations of city and citizenship, it is vital to come to terms with how the competitive, collaborative, and curative visions of Seattle relate and respond to one another. They do not represent a simple

3.1 The skyscrapers of livability.

historical trajectory, because the competitive global city remains dominant. Yet, while it is vital to address the ways in which the city and citizenship have been transformed by the pro-market boosters of competitive global development, the second and third sections of this chapter show that alternative articulations of the collaborative and curative global city are also embodied in Seattle's landscape for all who care to look for them and remember.

One place from which to see the three incarnations of Seattle's global status is the observation deck of the Space Needle. Had you been there on November 30, 1999, you would have seen the office towers, hotels, and condominiums of the competitive global city (fig. 3.1). But at the same time you would have seen the helicopters and even some of the tear-gas that police used against the Direct Action Network representatives as the union-led protestors massed and marched from an area beneath the Space Needle. It was the direct-action activists who so strengthened the association between Seattle and the global justice movement. They were the ones who chained themselves together at various sites around the Trade and Convention Center downtown, and who thereby ultimately succeeded in their goal of "shutting down the WTO." Yet, had it not been for the collaboration of vast crowds of other protestors in the big march coming south from the Space Needle—vast crowds "fifty thousand deep" in the haunting rap of Seattle's Blue Scholars—the police might never have been overwhelmed and the competitive business vision of a global Seattle would not have been so radically transformed.

In the pre-protest planning, the police and the mayor's office had imagined that most marchers would do a U-turn downtown and turn north, away from the convention center, away from confronting the WTO directly and back toward the Space Needle, where the crowds would disperse.[3] Yet while this plan to save the business-friendly city failed spectacularly in 1999, and while the collaboration between marchers and direct-action activists instead ensured Seattle's new reputation as a global city associated with global justice, the rerouting of the calls for justice,

democracy, and collaboration did, in a sense, come back north again to the South Lake Union area in the decade that followed. Thus, if you turn around to look north and east from the Space Needle today, an altogether different landscape becomes visible. Instead of the cosmopolitan crowds of Canadian, American, Mexican, European, and Asian protestors, you can now see signs of a different, more market-friendly cosmopolitanism. In the telling words of a 2009 *Seattle Times* article reporting on the Pacific Health Summit: "The glitterati of global health are gathering in Seattle," making the city the "Davos of global health."[4]

Most notably, near 500 Fifth Avenue North large displays advertising the work of the Bill and Melinda Gates Foundation picture people from all over the world as beneficiaries of global health philanthropy. Initiated in 1998, but officially founded a year after the Seattle protests in 2000, the foundation is now the largest (which is to say richest) private philanthropic institution in the world. Its new campus in South Lake Union, just a few blocks from the Space Needle, is set to be opened in the spring of 2011 (fig. 3.2) and is imagined by the foundation as a "hub for innovation and gatherings of experts from many fields, perspectives, and countries, who are dedicated to improving lives here and around the world." This work of innovation and gathering expertise is premised on the ethical axiom that, as the foundation's homepage says in capital letters, "ALL LIVES HAVE EQUAL VALUE." It is an ethical premise that clearly implies an inclusive and egalitarian conception of global citizenship, and there can be no doubting the foundation's core commitment to helping people everywhere in overcoming diverse forms of disease and dispossession. Moreover, the global health work that the foundation funds is not simply market fundamentalist, as critics sometimes suggest. Instead, its innovators see themselves as creating cures for "market failure." These cures are aimed at particular places and problems where capitalist marketmaking has been found wanting: for example, funding drug research for diseases that afflict the poor and that normally provide no profit motive for drug companies to get involved; or, organizing financial services for poor com-

3.2 The Gates Foundation campus under construction in South Lake Union.

munities shunned by big banks. Yet, as the managers at the Gates Foundation go about their business of funding these global market-curatives, their largesse also has considerable local labor and land market impacts, too.

Today's South Lake Union is being transformed by global philanthropy's vision of Seattle. Here, for example, you can find the imposing head offices of PATH, the Program for Appropriate Technology in Health (fig. 3.3), the largest single recipient of Gates funding (having received in the period up to 2007 forty-seven grants worth a total of $949 million, mostly for medical research).[5] And here, too, as we shall explore further below, you can find all sorts of other nongovernmental organizations and research centers that are also nonprofit in name but, like PATH, notably entrepreneurial, businesslike and market-oriented in nature. In short, while the landscape of South Lake Union may not have been the site of crowd dispersion that the

3.3 The new place of PATH.

Seattle police had hoped it would be in 1999, it has ultimately become a place where the anti-capitalist calls for global justice articulated in the WTO protests have been rerouted into more market-friendly activities today.

We will reexamine the complexities of the hybrid philanthro-capitalist landscape at the close of the chapter for the lessons it teaches about the degree to which city and citizenship in the age of curative global philanthropy diverge from the competitive and collaborative models of global Seattle that have come before. To begin with, though, we must first explore in more detail what the boosters' vision of the growth-oriented and globally competitive livable city actually involved, before turning to examine its repudiation in the streets in 1999.

The Competitive Global City:
Remaking Seattle as a World-Class Destination

"There is an extra perk that comes with living near the heart of downtown Seattle: global soul. A world-class city that hasn't lost sight of its down-home roots, Seattle offers urban living that trumps mere status with genuine international flavor." [6]

This glowing advertisement for the "global soul" that is supposedly available to condominium buyers in downtown Seattle no doubt sounds a little generic and, well, soulless. Whether you have read something similar in an in-flight magazine or in the travel section or new homes section of your own local newspaper, the familiarity of the sales pitch from one city to another may make you overlook the fact that such property market promotion has a local history. To get a sense of this history in Seattle, it is worth pausing before leaving the Space Needle to consider some other signs in the landscape that tell us something about the longer term development of urban boosterism in the city. Its strategies and styles have changed over time, but it has a relatively long history, going back at least as far as the Alaska–Yukon–Pacific Exposition of 1909: a world's fair aimed at showcasing Seattle as a city on the Pacific with

ties to colonial opportunities in the Philippines and Hawaii, as well as to the gold of the Klondike.

From the Space Needle you can still see the 1909 fairgrounds, now transformed into the main campus of the University of Washington. By contrast, the Space Needle itself, along with the monorail and the nearby grounds of the Seattle Center, still stands as a monument to another World's Fair, which in 1962 sought to promote Seattle quite differently as a center of science, industry, and innovation.

Trading on the importance of Boeing and associated industries in the space race, the Century 21 Exposition of 1962 revised the frontier themes of the 1909 World's Fair with a futuristic focus on Seattle's leadership in the advancement of American technology and planning.[7] The Soviet success with Sputnik I in 1957 had enabled the Seattle promoters of the fair to successfully lobby an anxious federal government for support, and thus the decidedly nationalistic and science-centric concerns of the space race meant that Century 21 turned into a classic Cold War concatenation of military-industrial capitalism, all presented as in the fun-loving interest of the liberal American citizen.[8]

Today, the Space Needle, public fountain, and KCTS Public TV studios still remain as reminders of the modernism and public inclusivity of this liberal American citizenship (as does the Elvis Presley movie *It Happened at the World's Fair*, which conveys the romance and fun of the exhibition along with a sense of trust in its public policing). Yet today, too, one can see more postmodern and privatized neoliberal landmarks (think *Frasier* instead of Elvis). These include: the plush make-over of the opera house; the nearby luxury condos; and, if you are out looking for a property to buy, the occasional sign emblazoned with the brand name—symbolizing the broader market take-over of the 1962 exposition themes—of what is now a real estate company called Century 21. Most conspicuous of all is the multicolored Experience Music Project (EMP) museum of rock and pop (fig. 3.4). There you can experience the affect and architecture of the transition from modern to postmodern urban boosterism in a place whose Web site tellingly boasts of being "a key economic driver among Seattle nonprofit arts and culture organizations . . . with $580 million dollars of local economic impact . . . [and] a spectacular, prominently visible structure [that] has the presence of a monumental sculpture." In all these ways, EMP illustrates how arts and culture are used in the crafting of "spectacular" world-class livability: a process that has gone hand in hand with the wider efforts to rebrand and rebuild Seattle in the era after the Boeing bust of the early 70s, after the Cold War, and after the related rise of postindustrial and postnational ideas about promoting individual cities in a globalized competition for investment and tourism.[9]

Created by the other Microsoft billionaire, Paul Allen, and designed by Frank Gehry, the EMP arguably continues the spirit of the 60s. But after paying the hefty entrance fee and putting on the headphone guides, visitors see and hear the rock and pop music as an assemblage of individualized and objectified museum experiences cut off from the shared sounds and communal vibes of the concert crowd. The outcome undoubtedly makes American rock more accessible and understandable for uninitiated visitors, but it also repackages the music's popular energy and occasional notes of radicalism in ways that diminish their social message. What is left instead seems to be just personalized cultural improvement tied to competitive global city promotion. Advertised as a world-class tourist destination, the EMP would therefore seem to co-opt and commodify the liberalism of sixties Seattle, transforming it into the same sort of neoliberal experience in individualized edification available to globe-trotting museumgoers everywhere, Gehry's Guggenheim museum in Bilbao included.[10] And whether Gehry calls himself a "deconstructivist" architect or not, the external architecture of the EMP building would similarly seem to deconstruct the changing meaning of citizenship in the world-class livable city: its fanciful outer skin swallowing up paying passengers who arrive on the mass-transit model turned tourist attraction that is today's monorail.

The monorail's changing symbolic significance notwithstanding, you can still take it back out of the EMP shell. If you do, the journey south toward the downtown

3.4 The Experience Music Project (EMP).

brings many more landmarks of the globally competitive livable city into view. Through the Belltown neighborhood, new luxury condominiums are advertised to the left and right, nearly always, it seems, using images of swingers in sexy clothing or fitness gear to lure would-be buyers into the latest "hot" new urban lifestyle (read, real estate) opportunity. Then, after getting off the monorail at Westlake Center, the commercial heart of the livable city beckons and beats on all sides. Here are the brand-name shops: Abercrombie and Fitch, Anthropologie, Banana Republic, Niketown, Nordstrom, and Planet Hollywood. Here close by, too, is Pike Place Market, which, like Boston's Faneuil Hall and other destination urban markets the world over, is now so full of tourists buying trinkets and T-shirts that it is hard to imagine how the vegetable stalls ever do a profitable business. And here also, not much more than a salmon toss from the Market, are many of Seattle's famous finedining restaurants: Flying Fish, Campagne, Place Pigalle, Maximilien, the Dahlia Lounge, and Wild Ginger. Benaroya Hall, another asset in the livable-city portfolio, welcomes Seattle Symphony audiences into an atrium topped with Dale Chihuly glasswork. And just another block down Union—a street name that does not mesh well with the surrounding spectacle of postindustrial redevelopment—

the Seattle Art Museum offers its own good taste to the wrap-around picture of urbane livability.

While the picture may be pretty, the way it was painted was not. Securing "the spectacular city" instead involved the diversion of public funds into heavily subsidized commercial ventures, along with all sorts of authoritarian efforts to rid the downtown of the downtrodden and homeless.[11] With the economic challenges facing the city after the busts of the 1970s and '80s, local leaders, such as mayors Norm Rice and Paul Schell, were pressured by the same worldwide economic forces that were imposing business-friendly norms of competitive urban investment and boosterism right across North America. At the same time, city elites saw the project of attracting the wealthy to downtown as simultaneously necessitating the marginalization of the urban poor. In other words, the project of revitalization also became about making citizenship in the city more exclusive. One reason why, it seems, was a chain of associations that linked the poor with urban legends about the decay and death of cities, associations that at the same time linked the wealthy with visions of vitality, livability, and global success. On the one side, there was the fear of decline—as with Detroit, Buffalo, and other cities in the U.S. rust belt. On the other side, there was the hope of winning in the international competition for investment like other competitive West Coast cities, such as San Francisco and Vancouver, B.C. Thus, urban governance was understood to be a bifurcated choice between industrial-era managerialism associated with supporting citizens of all classes (albeit unequally) and global-era entrepreneurialism associated with the gentrification of urban citizenship and the punitive policing and exclusion of the poor.

The shift toward more exclusive and authoritarian ideas about citizenship in Seattle also worked as something of a distraction from the investment of public money in all the new commercial and cultural ventures. These big public subsidies included millions of dollars for two new sports stadiums, and the fact that these stadiums still ended up being named after big private corporations was a good indication of who the real winners were in the entrepreneurial

pursuit of global city competitiveness. However, and this is key to how citizenship in the livable city became increasingly exclusive, instead of being made to feel like losers themselves, the tax-paying public was instead encouraged to concern itself with the threats posed to security by a loser class of poor people. Following a pattern that is found in cities all around the globe,[12] extensive efforts were made to reassure urban and suburban shoppers, visiting tourists, and professional office workers that the beggars would be banished. The livable city of the elites had to become unlivable for others.

Not all Seattle art lovers and concertgoers see the poor and homeless as noncitizens without rights to the city. It is true that some notable individuals and editorialists pontificated as if they truly believed this. An aggressive city attorney named Mark Sidran, for example, did his best to look like the Rudy Giuliani of Seattle by pushing a set of anti-vagrancy ordinances through the City Council in 1993, and these and other allied initiatives by city elites were welcomed by the ever-worried *Seattle Times* editorial page. "Seattleites have made huge investments to make the downtown an economically viable, physically inviting place," the editors explained. "Those achievements are threatened when some streets and parks become unpoliced havens for panhandlers and unruly drunks."[13] Likewise, the language later used in 1997 to defeat plans to build a service center for the homeless on Third Avenue near Benaroya Hall was littered with similar depictions of the poor as uncivil and unsavory subcitizens. "It's like [putting] a meat rendering plant next to a high-income residential area," argued one opponent of the service center. "It's a safety thing," explained another, "I might get accosted, I might get beat up, I might get robbed, I might get raped, or whatever."[14]

Structurally, the actual outcome of all the revitalization efforts, anti-vagrancy laws, and associated policing policies was to make urban citizenship more and more a matter of class privilege. In King County as a whole, growing income inequality shows up only as a small increase in the Gini coefficient from 0.144 in 1970 to 0.187 in 2000.

However, in Seattle's downtown itself, the gentrification of urban citizenship was much more marked in increasing income inequality, increasing per-household income (which, in adjusted 2007 dollars, went from $28,669 in 1980 to $42,562 in 1990 to $63,088 in 2000), and in the imagination and enforcement of who belonged and who did not.[15] The liberal American inclusiveness planned for Century 21 was replaced on the eve of the real twenty-first century by a neoliberal market-sorting that undermined the universality of an urban future for all. As John Fox and John Reese of the Seattle Displacement Coalition predicted in their campaign against Sidran's ordinances: "When the civil rights of the poor and homeless are at stake, the rights of all of us are threatened."[16]

In terms of global city developments, one other aspect of the changing class composition of citizenship in Seattle deserves attention: the ways in which the global business classes became interwoven with the efforts to package and sell the Seattle scene internationally. While the poor and homeless were being banished and diminished as rights-bearing citizens, all sorts of efforts were afoot to make Seattle part of a global network of "gateway" cities, in which borderless business-class citizens and allied professional and consumer classes might enjoy new rights of membership and mobility transnationally. The binational, cross-border concept of Cascadia, for example, was promoted all through the 1990s and right up to the 2010 Winter Olympics in Vancouver and Whistler in the hope that, by bonding with British Columbia, Washington State (and Seattle) could attract more interest from cosmopolitan capitalists. The idealized Cascadian citizens as conceptualized by these binational boosters were not environmental activists (as in the original bioregional conceptualization of Cascadia), but rather capital-carrying global investors and tourists.[17] This is reflected in the denationalization of citizenship in global cities more generally. Rights to buy property and make contracts, to move freely with work visas, or simply to enjoy fusion cuisine in other cultures can all be understood as developments to denationalize citizenship. The result is that the livability of the American global city

is opened and sold to foreigners with capital, even as poor Americans (including, for example, many Native Americans in Seattle) are effectively excluded from full citizenship.

If we return to our tour of Seattle, the trip from the Space Needle south through the downtown leads us eventually to a site where the cheek-by-jowl juxtapositions of global-city citizenship and its excluded "others" (remember Benaroya Hall and the meat-rendering plant) are especially marked in the landscape. Walk between Pioneer Square, the International District, and the areas south of downtown called SoDo, and you will see signs of this extraordinary intersection on every side (SoDo itself suggests Soho/NewYork/London global city exclusivity). There are foreign tourists and city shoppers in the Pioneer Square coffee shops. Around the corner, to the chagrin of the local shop owners, are the homeless in Occidental Park and the hungry waiting for the Seattle Union Gospel Mission to open. Down the road on game days, sports fans head to their class-assigned positions in the stadiums (which accommodate more luxury corporate boxes than the Kingdome they replaced). If there is no game, there are still the business-day lines of professional staff heading into the headquarters of Starbucks, some returning from overseas trips for the company. Yet right by Starbucks is the SoDo Home Depot, with its all-American indication of the curtailment of citizenship for working-class cosmopolitans: Latino laborers lined up in hopes of being picked for work on local building sites. And right behind Home Depot and the Starbucks offices is a different kind of line: containers packed with the products of Asian laborers, who are themselves managed by far-flung, subcitizenship systems such as the *hokou* registration regime in China. In the summer, there are also cruise ships. Packed with retirees enjoying the last leisure opportunities afforded by diminishing social security, these excursions, again, tell a story of citizenship becoming more exclusive.

Back closer to downtown is the International District. Once a euphemism for Chinatown, the name now signifies the bigger story of Pacific Rim development and gentrification; the upscale Uwajimaya superstore and apartments, for example, demonstrate the ways in which Seattle's Asian ties now bring wealth and investment to the city, even if the old rooming houses and cheap Chinese restaurants still share space in the district.

In all these ways, the Seattle landscape illustrates how class—something that most Seattle residents tend not to talk about—influences who wins and who loses amidst global city development and denationalization.

Capital, at least in the form of American dollars, introduces another leitmotif in the selling of Seattle as a competitive global city: namely, greenery. "Seattle, the Emerald City, is the jewel of the Northwest, the queen of the Evergreen State, the many-faceted city of space, elegance, magic and beauty," or so claimed the Seattle–King County Convention and Visitors Bureau back in the 1980s. Taken from *The Wonderful Wizard of Oz*, though, the name Emerald City also said something important about the eco-fashioning of the global city. "Just to amuse myself," explained Oz to Dorothy in the original story, "I ordered them to build this City, and my palace. Then I thought . . . I would call it the Emerald City, and to make the name fit better I put green spectacles on all the people, so that everything they saw was green." As Matthew Klingle outlines in his eloquent account of Seattle's environmental history, this allusion to the wizardry of Oz is telling because myths of Seattle being closer to nature than other cities rest on similar spectacular illusions.[18] Too much of the green spectacle, it turns out, is like wearing green spectacles: the decline of Seattle's salmon, for instance, is often obscured by the widespread use of salmon signs and symbols in the visual branding of the city.

Notwithstanding other ongoing environmental problems, such as the Duwamish Superfund sites, the worsening of Puget Sound's water quality, and the death of most first-born Orca whales because of toxins that collect in their mothers' milk, Klingle ends his own story hopefully, suggesting that at least a few of their green illusions may now lead Seattleites toward more genuinely sustainable models of development. Whatever we may make of this, there can be no doubting that along with all the sustainable develop-

ment discourse in the city, a great deal of what has actually been sustained is the idea of world-class livability itself. Again, the question of class—specifically, which classes can see the green vision and partake of its rewards—needs answering. In 2009 *Outside Magazine* named Seattle the second-best U.S. city.[19] The fact that Seattle was ranked so high no doubt gave local promoters of the competitive global-city vision another good reason to raise their glasses of Washington State reserve wine. But when one reflects on what was also factored into the ranking by the magazine— including "percentage of college degrees" and "income level," as well as "quality and proximity to biking, running, paddling, hiking and skiing"—the exclusivity of the vision becomes clear. Seeing green, like buying ski-tickets, clearly costs, making the "outside" Emerald City livability features inaccessible to class outsiders.

In 2006 a global-city newspaper report answered the class question with unusual candor.[20] The *Seattle Post-Intelligencer* noted that the Urban Land Institute/Price Waterhouse rankings of that year had rated Seattle the "top global gateway city" in the United States, and it acknowledged that high-ranking cities were "expensive and elitist," because "people pay a premium to live in them." Such exclusivity was fine, though, the report reassured readers, because it created a virtuous cycle of the real green stuff of consequence, namely dollars: "Companies pay handsomely for brainpower," it went on, "so employees can afford to live in these places. In turn, an affluent population supports the art community, which drives tourism." It is this neat ecosystem of profit-making that repeatedly turns sustainable development into sustaining competitive capitalist growth in the global city. It also allows boosters to go on selling the city as a great green place to live. But more than this, it has also allowed businesses to turn Seattle and its image of livability into a tool for selling other things, including everything from outdoor sports equipment and clothing to cars and SUVs advertised on downtown buildings and skyline.

Most notably, the imagery of the Emerald City has become a way of globally selling another brand with green signage: Starbucks. As James Lyons explains in his study of the Seattle-coffee connection, "retailers [such as Starbucks] sought to cultivate and mobilize associations with Seattle, and make explicit and oblique signifiers of the city crucial to their promotional and branding activities."[21] In this way, Seattle effectively reached competitive global city nirvana: its highly ranked livability helped sell a product associated with class distinction and good taste in many other places around the world.

Given the coffee connection to the Emerald City's global branding, it is notable that since the financial crises of 2008–9, Starbucks has responded locally to falling demand by rebranding and shedding its green signs. The new strategy converts some of its coffee shops back into nonbrand local coffee shops.[22] Might this reflect the falling stature of the Emerald City brand, too? After all, the financial crisis hit Seattle hard: emptying downtown offices, shops, and condos; lowering its gateway-city ranking, and spectacularly bankrupting the biggest Seattle-based bank, Washington Mutual. Here again, what is impressive about the promoters is their repeated return to the resources they see in the local landscape, always implying that its sublime physical geography somehow ensures Seattle's success in the ongoing race to develop a globally competitive economic geography. Thus, in 2010 the Greater Seattle Chamber of Commerce continued to promote the city with the following sound bites:

Ringed by snow-capped mountain peaks.

Crisscrossed by fresh- and salt-waterways.

Home to urban centers, small cities and vibrant neighborhoods.

Greater Seattle is an incredible place to live as well as to do business.

The continuing growth of the Puget Sound region is a clear indication of its desirability as a place to do business and to live.[23]

What remains less remarked upon, but nonetheless is very noticeable in all the promotional idealization of the city, is the idea that its citizens should also be considered special and exclusive in terms of class. Education is key here, and the boosters are keen to underline that: "Residents are highly educated; in fact, Seattle consistently ranks at the top in national polls for college degrees per capita." But important, too, are taste and income. Victories for local wineries are listed alongside all the "best city," "fast city," and "gateway city" successes. The high median household income for King County was projected at $67,338 in February 2010, and the Trade and Development Alliance states that "the Puget Sound region is a relatively affluent area." Whether this observation is a discreet acknowledgment or a bragging boast, the underlying point about class is surely clear. Seattle is a city where wealthy transnational investors and tourists are warmly welcomed into the livability enjoyed by similarly privileged local citizens. It is a gateway city, in other words, that puts class exclusivity into the term "world class," redefining belonging, ownership, and the mobility rights of citizenship in ways that are commensurate with the competition for global capital. In 1999 it was this same eagerness to embrace the leaders of corporate globalization that led the city to play host to the WTO. However, as we shall now examine in more detail, this invitation to the borderless business class cost the competitive global-city promoters dearly.

The Collaborative Global City: Retaking Seattle through World Class Resistance

"It is time to raise the social and political cost to those who aim to increase the destruction and misery caused by corporate globalization."[24]

So began the invitation to "Come to Seattle" by the Direct Action Network as it reached out to diverse groups, welcoming them openly to collaborate in staging the anti-WTO protests. "This will help catalyze desperately needed mass movements," it went on, "capable of challenging global capital and making radical change and social revolution." Clearly, this invitation was very different in tone and goals from those prepared by global-city proponents (a contrast also addressed in chapter 7). Before the protests, for instance, Mayor Paul Schell had depicted the WTO ministerial as a "momentous, exciting affair for Seattle." Going on to make the classic claims of a booster for world-class status, the mayor argued that hosting the event spoke "to the growing stature of Seattle's place on the world stage."[25]

Instead of joining the competition for global capital, the Direct Action Network's invitation spoke to how such competition disciplines and diminishes citizenship and democracy. And, instead of class exclusivity, it called for class struggle and other struggles for global justice to be joined in shared nonviolent demonstrations in the streets. Thus, along with issuing the challenge to global capital, the protestors planned for a notably collaborative gathering that would be inclusive and radically democratic in its organization. These plans for the protests—which were extensive and involved all sorts of educational efforts— ensured, in turn, that the collaborations that ensued were multiple and far from just a singular, spontaneous outcry of the multitude.[26]

The political philosophers Michael Hardt and Antonio Negri are no doubt right to claim that "the real importance of Seattle was to provide a 'convergence center' for all the grievances against the global system."[27] But such convergence had to be invited and planned, and this is where real on-the-ground collaboration became so important. At least three collaborations stand out for the ways in which they sutured together the reputation of Seattle as a global city and the idea of global justice. The first was the transnational cooperation of local and global activists sharing common grievances against corporate globalization and WTO rule-making. The second was the "Teamsters and Turtles" collaboration between the unionized working class and various environmental movements that joined the protests. And the third was what might be called the "No Globalization without Representation" collaboration between the direct action activists, who did the bulk of the orga-

nizing and planning, and the less radical, but nevertheless vast and engaged population of ordinary citizens for whom the reworking of the old revolutionary adage about taxation and representation resonated increasingly as the struggle over representational rights unfolded in and over the public space of the city. By examining each of these three collaborations in turn, it is possible to reflect on how they transformed the meaning of world-class citizenship by revising and re-visioning its urban geography.

The first collaboration, between local and global protestors, is important because one of the main mistakes made in media coverage was to ignore the "globalization from below" that the protests represented by simply dubbing them "anti-globalization." The protestations were instead aimed at a very particular pro-privatization, pro-deregulation, pro-corporate approach to WTO-enforced globalization. A more nuanced and historically savvy description of this critical political position is "anti-neo-liberalism." But such nuance, along with the fact that the protests involved global planning, global participation, and global solidarity, could be negated by the "anti-globalization" label. This negation was easy to do from afar, not least because the involvement of a few right-wing, xenophobic, and anti-trade crusaders (such as Patrick Buchanan) could convince pro-business commentators and economists to conclude that all the protestors were simply old-fashioned "protectionists." But for those who witnessed or joined the protests on the streets of Seattle, an altogether more global set of arguments and alliances could be seen and heard, advancing a new, postnationalistc protectionism.

The old union chant, "An Injury to One Is an Injury to All," reverberated loudly down Fifth Avenue in the march led by the AFL-CIO, but it was given new, postnational meaning by the attendant signs calling for a *global* struggle for a better, more humane, environmentally friendly, and democratically accountable approach to globalization (fig. 3.5). "The End of Suffering Comes about Locally through Global Solidarity," read one. "No Child Labor," said others. And "Free—to exploit people and nature—Trade" explained one of the spectral street puppets: making visible the lack of freedoms for people and nature that haunt WTO-enforced free trade. If this was protectionism, therefore, it was based on a new postnational and nonchauvinistic desire to protect people and the environment *everywhere*. And further emphasizing both the promise and practicality of this new kind of protectionism was the profoundly globalized collaboration of local and global protestors in the organization, implementation, and celebration of the protest itself. As Anuradha Mittal has subsequently put it: "The World Turned Out in Seattle."[28]

While the Direct Action Network was sending out invitations ahead of the protests, another group, the International Forum on Globalization (IFG) based in San Francisco, organized an extraordinary teach-in on the WTO in Benaroya Hall. This free event generated such interest that it was virtually impossible to get in without reserved passes (a problem with access based on over popularity that was in telling contrast to the problem of access at the WTO meetings in subsequent days). The teach-in featured diverse speakers (including Vandana Shiva and a number of other foreign experts), who delivered powerful if pedagogic speeches explaining the destructive downward harmonization unleashed by WTO rule-making: the prohibition of seed-saving practices by TRIPS; the "race to the bottom" dynamic created by liberalized international competition; and the direct dismantling of environmental, labor, and community protections as a result of WTO nontariff barrier removal laws. These are all now lessons about legal neoliberalism routinely taught in university classes on globalization, but back in 1999 they arrived as insights from activists working around the world on the global impact of WTO law.

Carrying their lessons on corporate globalization into the streets on November 30, 1999, the IFG and associated international trade law activists went on to mingle with other cross-border organizers concerned with migrant rights, indigenous rights, and human rights. Anti-war signs, Canadian maple leaves, and Mexican flags were all abundantly evident, as too were all sorts of foreign-language signs held aloft by small contingents of protestors from Germany, France, Korea, South Africa, and the

Caribbean. Moreover, many banners and placards held by these groups made it clear that it was precisely the cross-border character of their organizing that made it make sense for them to be there together in Seattle. Maquildora women's solidarity organizers were there next to "FIX IT OR NIX IT" Canadian organizers; the Southwest Network for Environmental and Economic Justice held up a banner saying "BUILDING POWER WITHOUT BORDERS IN THE SPIRT OF OUR PEOPLE"; and a pro-Cuban group walked next to Chicanos and Chicanas carrying a sign saying "SOMOS UNO PORQUE AMERICA ES UNO."

Joined together in the streets of Seattle, it was this polyglot crew of world citizens making their local-turned-global cases for another kind of globalization that then went on to merge with the direct-action activists battling with the Seattle police around the Trade and Convention Center. The merging did not go smoothly, and it would be wrong to suggest that the overall collaborative effect had the character of a centrally planned and well-coordinated action. In fact, union plans enforced by marshals wearing orange hats succeeded in steering many protestors away from the direction of the barricades and police violence.

Meanwhile, that violence combined with the opportunistic defacement of shop fronts by Black Block vandals, made many marchers nervous about going closer to the action. Odd scenes in which anarchy mixed with fascination and bewilderment were common. Yet, there was solidarity and a sense of shared citizenship amid the confusion.[29] And, in the end, so many marching protestors filled the downtown streets that crowd control became impossible for the police. Running out of tear gas, they therefore had to abandon their initial efforts to clear the streets. As a result, WTO delegates found it next to impossible to leave their hotels, and even members of President Clinton's administration complained to the Secret Service that they could not get to the Trade and Convention Center. The collaborative global city had made competitive global city winners into losers.[30]

In showing that street protests in a world-class American city could produce meaningful resistance to global

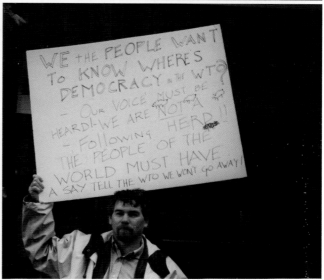

3.5 Global solidarity and resistance during the WTO protests in Seattle.

capitalism, the protestors also succeeded in making the name of Seattle an emblem of anti-neoliberal organizing across the globe. This, though, was not just about making Seattle a single signifier of global resistance. As significant as this semiotic association with anti-neoliberalism remained afterwards, there were also more material linkages with real-world struggles for global justice as well. As one of the organizers summarized for fellow activists in the days that followed, global collaborations with Seattle were real and happened in real time.

People across the globe took action in solidarity. In India thousands of farmers in Karnataka marched to Bangalore, and over a thousand villagers from Anjar in Narmada Valley held a procession. Thousands took to the streets in the Philippines, Portugal, Pakistan, Turkey, Korea, and across Europe, the United States and Canada. 75,000 people marched in 80 different French cities and 800 miners clashed with police.[31]

After 1999, what many global activists celebrated as "the spirit of Seattle" lived on and grew. The next WTO meeting took place in Doha in 2001, where active civil society engagement was effectively prohibited, but two years later in 2003, when the ministerial was held in Cancun protestors again pressed their critical cases, and the same happened again in Hong Kong in 2005. Over this time, campaigns against other institutions of global corporate governance—ranging from the World Bank and IMF to the G8 and G20—also continued to attract widespread "Seattle-style" opposition. In terms of citizenship, these subsequent developments meant that the collaborative global connections exemplified by Seattle went on to inspire a form of democratic denationalization involving civil society groups in other newly collaborative global cities across the planet. Prague, Gothenburg, Montreal, Mumbai, Genoa, Cape Town, Nice, New York, and even Washington, D.C., have all come to join this list. It is true that the policing of this new, global-city resistance has also become routinely draconian and preemptive in its zoning and targeting of protestors. In becoming more authoritarian, however, the policing reveals that the market-makers of global capital continue to be haunted by the specter of Seattle. In this sense, the ongoing repression of the global justice movement in other collaborative global cities further vindicates Seattle critics, who argued that the police reprisals made manifest the repressive unfreedom at the heart of global free-market rule-making. While academics and jurists may well go on debating what sort of spaces democratic states should provide for protest within particular national jurisdictions, the larger lessons spiraling out of Seattle relate to demands for a wider, denationalized, collaborative freedom across the planetary space of global citizenship itself. This then is no doubt why political philosophers conclude that "Seattle was the first global protest: the first major protest against the global system as a whole, the first real convergence of the innumerable grievances against the injustices and inequalities of the global system."[32]

The second collaboration that further ensured Seattle's global salience and transnational ties was the coming together of the "Teamsters and Turtles." The sense of joy in this collaboration was evident on the streets. Environmentalists marched happily alongside rank-and-file union members, the two groups clearly moving beyond divisive stereotypes of one another. In terms of citizenship, the Teamsters and Turtles alliance also represented a reimagination of the space of democracy: the global scope of environmental concerns helping at least at a symbolic level to denationalize union concerns about outsourcing and reawaken the old internationalist spirit made so famous by the International Workers of the World in Seattle in the early twentieth century. Speech makers from both the environmental movement and the unions articulated a shared global perspective on improving both environmental and labor standards globally. To be sure, some union placards still railed against the offshore siting of jobs. And some environmentalists still spoke of the need for environmental protections, without addressing the plight of workers. But on November 30, 1999, these ongoing tensions were replaced by a shared global concern articulated loudly with linked arms in Seattle streets.

Capturing the collaborative qualities of the protests in their own philosophical style, Hardt and Negri also emphasize that "the unexpected collaboration of trade unionists and environmentalists was just the tip of the iceberg." The protests "brought together innumerable other groups expressing their grievances against the global system—those against the practices of huge agribusiness corporations, those against the prison system, those against the crushing debt of African countries, those against IMF controls of national economic policies . . . and so on, ad infinitum."[33] The Teamsters and Turtles alliteration stood for a much longer chain of allied social movement solidarities. It was for the same reason, though, that less sympathetic commentators poured scorn on the collaborations as sixties-style incoherence. Corporate globalization guru Thomas Friedman dismissed the protests in this way as "Senseless in Seattle": "Is there anything more ridiculous in the news today then the protests against the World Trade Organization in Seattle? These anti-W.T.O. protesters—who are a Noah's ark of flat-earth advocates, protectionist trade unions and yuppies looking for their 1960's fix—are protesting against the wrong target with the wrong tools."[34] Given that Friedman subsequently went on to write a book that used the flat-earth metaphor to represent the world-changing implications of globalization, this was an especially egregious, and self-incriminating, effort at ridicule.[35] The protestors were not out-of-touch flat-earthers at all, nor did they demand or represent a return to the American-centric anti-war radicalism of the 1960s. Instead, their transnational and trans-grievance collaboration in united criticism of the WTO effectively anticipated and argued against precisely the flattened vision of the world that Friedman himself went on to peddle in his best-selling book. In doing so, it also involved a third collaboration: a collaboration that used the tools of American democracy to argue for more global democracy in institutions of global governance such as the WTO.

NO GLOBALIZATION WITHOUT REPRESENTATION was probably one of the most popular and powerful slogans of the whole protest. It was used by Sierra Club environmentalists and union protestors alike, and it clearly resonated with the wide variety of ordinary Seattleites who, at Benaroya Hall and elsewhere, clamored against the unelected, unaccountable, and generally undemocratic nature of WTO administration. As the Sierra Club sign made obvious, the slogan deliberately recalled the spirit of democratic uprising associated with the Revolutionary War. And out in the harbor, a protest stunt by the steelmakers union dumped fake steel bars from Asia in the water, in a reference to the Boston Tea Party. But the slogan itself—No Globalization without Representation—also departed from the original nationalism of "No Taxation without Representation" (not to mention the xenophobia of today's tea-partiers). By substituting "globalization" for the American concern with "taxation," it pointed postnationally toward a vision of global citizenship, to the idea that economic globalization had to come with the globalization of democracy. The primary target here was the WTO for not allowing any stakeholders except trade lawyers and trade ministers into meetings over trade complaints. But the larger argument, and one made powerfully and poetically in other signs recalling and revising U.S. constitutionalism, was that ordinary people need new ways to make global institutions more democratically accountable.

As the police confrontations became more violent and more people were arrested and imprisoned, a civil emergency was declared. City streets were cordoned off and the National Guard was brought in. In this context, the importance of demanding democratic rights became increasingly resonant locally as well as globally. Ordinary members of the Seattle public who joined the protests, and even just passersby, were arrested along with the direct-action activists. Residents of the Capitol Hill neighborhood where officers chased demonstrators and fired tear gas also experienced police violence and arrests. And those who were arrested went on to face further abuse and detention without access to legal counsel. As Jill Friedberg's brilliant movie *This is What Democracy Looks Like* documents in detail, all of these developments only increased solidarity among citizens, marchers, and activists. Outside King County jail, a huge vigil took place, and a chant that had already been heard all day reverberated into the night: "Whose Streets? Our Streets!" These and other similar moments illustrated that the demands for global democracy, such as those made by the Sierra Club and union protest signs, were suddenly, and for many people surprisingly, salient for Seattle citizens, too. Global justice had also come to be about local justice in the collaborative global city.

The Curative Global City: Remaking Seattle as a World-Class Philanthropy Center

"Every year, millions of people in developing countries die from diseases, including malaria and tuberculosis that have been all but forgotten in rich countries. For these diseases, the economics of the marketplace are not sufficient to commercially justify the large-scale investment needed to develop and deliver vaccines and drugs. Through global advocacy, the Bill & Melinda Gates Foundation is working to address this market failure by promoting innovative health financing mechanisms that provide better incentives to the private sector to create global public goods." [36]

Providing here a frank acknowledgment of global health inequalities, explicitly referring to markets not serving the poor, and making references to "market failure" and the "failures of the marketplace," Joe Cerell's explanation of how the Gates Foundation approaches its work sounds a little like some of the arguments made against the WTO in Seattle in 1999. Yet published on the Web site of the International Monetary Fund and emphasizing public-private partnerships as the correct cure for market failure, there are very big differences, too. The concern with global injustice and the desire to honor the idea that all lives are of equal value still obviously animate this vision of global health philanthropy. In this sense, the basic concept of global citizenship shares something with the inclusive impulses

of Seattle in 1999. But, at the same time, democratic collaboration, support for governments resisting structural adjustment policies, and demands for popular protest and public accountability are not seen as ways to win global justice. Instead, the Gates Foundation focuses on strategies deliberately designed to cure market failure using private-sector incentive schemes and market-mediated innovations in global health, development, and education. Market fundamentalism, business-knows-best ideology, and an absolute faith in the power of markets are thus not radically challenged as they were in 1999. Instead, they have been revised and replaced by what is better termed "market foster-care" through diverse plans for micro-management and development.

Huge debates rage online between ardent critics (represented, for instance, by gateskeepers.civiblog.org) and sympathetic observers (represented by Matthew Bishop and Michael Green's philanthro-capitalism.net) of the impact Seattle philanthropy is having on global health, development, and education. Here the much more specific aim is to ask how the philanthro-capitalist cures are related to Seattle as a global city and the denationalization of citizenship. The strong ethical insistence on the equal value of all human life in the statements of the Gates Foundation translates into a wider commitment by much of the rest of the Seattle-based global philanthropy community into seeing life across all national borders as worth saving. The concept of citizenship underpinning their efforts is in this sense global and denationalized. At the same time, a closer look at the details of the programs envisioned and implemented by Seattle's philanthro-capitalists indicates that this same global inclusivity is also radically individualizing. The goals are about empowering individuals—not nation-states—to overcome disease and diminished circumstances where markets have failed. Thus, just as activists' critiques of market failure are replaced by an emphasis on micro-market management, so, too, it seems, are the protestors' ideas of collaborative global citizenship and collective responsibility replaced by an emphasis on individual empowerment and personal responsibility.

Along with the individualizing imperatives comes a notably entrepreneurial reimagination of the geography of intervention. Target areas for global philanthropic involvement are imagined and mapped as investment sites where the philanthro-capitalists seek to create the greatest possible return on their money. To be sure, the return is measured in terms of improved health outcomes, environmental enhancements, and human capital development rather than on purely monetary measures of capital. Yet, the broad investment vision and vocabulary, the talk of more health for money, and all the associated investments in innovations ranging from vaccines to transgenic seeds to water contamination measurement tools would seem to suggest that Seattle's philanthro-capitalists bring some of the same business outlook to their approach to global empowerment as the city's competitive boosters employ to advocate for local investment. What changes is that the investment logic is inverted: going outward from Seattle to the rest of the world and enfolding it in a reconceptualization of planetary space as the staging ground for Seattle's global ambition. This approach is no doubt much healthier than the spread of Starbucks, but there remains nonetheless a certain sort of promotional logic at work that is very much of a piece with the ranking boasts that preoccupy Seattle's more traditional urban boosters. The result, not surprisingly, represents a notable hybridization of the meaning of world class.

One example of the hybrid reconceptualization of world-class Seattle, produced by leaders of the curative global city, has been cartographic. Created by the Washington Global Health Alliance, it takes the form of world maps that represent all the other parts of the planet to which the global curative efforts of Washington State, and Seattle-based institutions in particular, are linked by the integrative ties and investments of global health development (fig. 3.6).[37] The Washington Global Health Alliance's global vision clearly represents something different and distinctly denationalized. Scaling up John Snow's inaugural cartography of public health citizenship in Victorian London, it might even be argued to represent a new "ghost map" of

global public health citizenship.[38] But investment-oriented and promotional as it is—a form of mapping that would register cholera deaths as a new opportunity for targeted curative capacity-building rather than as an indication of an infected water source—this denationalized remapping of health citizenship clearly has a local agenda. In short, it represents an attempt to leverage local economic development opportunities out of the global developments being planned by Seattle's philanthro-capitalists. As a result, Washington Global Health Alliance meetings have featured Washington State's senators and governors celebrating what they see as the local job and growth opportunities tied to global health, and reports from these gatherings help highlight the ways in which the curative global-city vision emerges from the ongoing tension between competitive and collaborative ideas about Seattle's ties to global health.

"Global health is now part of our region's economy," explained Senator Patty Murray at one joint Health Alliance and Chamber of Commerce meeting in 2008.[39] Her colleague, Senator Maria Cantwell, added that the economic impact included an extra $4 billion in business activities in the state. The reporter commenting on the resulting big business interest in this emergent "industry" further captured the way in which enduring concerns with the competitive global city rankings race were overshadowing the proceedings. "The field of global health, once perhaps regarded as a noble humanitarian endeavor aimed at healing and helping the world's poorest people," he began, "is now becoming something of an 'emerging industry' that the Seattle business community sees as a highly competitive enterprise that is ours to lose."

The same reporter went on in his article to highlight more collaborative concepts of the region's role in global health still being articulated at the conference. "The bottom line here is jobs and income for us as opposed to being focused on assisting poor people in the developing world," argued Dr. Steve Gloyd, the head of Health Alliance International (HAI) and director for global health curriculum at the University of Washington. Joining him in resisting all the corporate pressures, Wendy Johnson, another UW and HAI

THE STATE'S GLOBAL HEALTH ORGANIZATIONS ARE GROWING – BRINGING THE BENEFITS OF INNOVATIVE RESEARCH, EDUCATION, TRAINING, AND PROGRAM DELIVERY TO PEOPLE ACROSS THE WORLD

WGHA organizations have a global presence. The map below shows the international locations of WGHA organizations' offices and facilities. WGHA organizations operate offices and lab facilities in 29 countries and 58 cities worldwide. This does not include the hundreds of project sites and the use of partner facilities.

	FULL-TIME EQUIVALENTS, IN WASHINGTON	2,324
	FULL-TIME EQUIVALENTS, OUTSIDE OF WASHINGTON	2,150
	FULL-TIME EQUIVALENTS, TOTAL	4,474

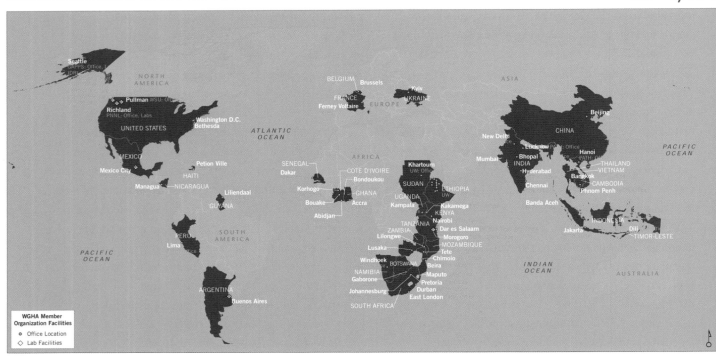

3.6 Curative global capacity building.

Source: Washington Global Health Alliance.

doctor, said: "I think it's a slippery slope. Once you define this as an industry, you tend to start investing in the kinds of projects that support jobs and programs rather than in what's really needed to solve health inequities in poor countries."

What makes these tensions over the meaning of global health in Seattle especially significant is that they help explain how the curative philanthro-capitalist vision sees things in relation to the tension. The same reporter concluded his article by quoting Cheryl Scott, the conference co-chair and a senior adviser at the Gates Foundation. She told him that "she understood the concern about treating global health as if it were just another commercial enterprise." However, she explained, this was largely a semantic tool aimed at arriving at a "framework" for building a unique kind of public-private partnership. "None of this is to imply that we (the Gates Foundation) think global health can work solely as a private model," she said. Contrary to the criticism that the Seattle philanthropy favors privatizing global health, Scott said that many of their projects are being done "in close collaboration with Third World governments." With ongoing pressure from global public health advocates who emphasize strengthening health systems, and with their own rigorous commitment to objective results assessment, it is no doubt true that the Gates Foundation and other allied institutions in the city will continue to build global collaborations. Indeed, with new critical studies published in medical journals posing questions about "the geographical location of primary recipients [of funding]," the pressure to share the wealth outside of Seattle looks set to rise.[40] As they go global, though, Seattle's philanthro-capitalists also continue to build new buildings in South Lake Union, and here the pull on curative global city ideas seems very much more in the direction of the competitive global city with which we began.

One sign of the times that indicates which way global Seattle is headed is that the Seattle Displacement Coalition, which fought for the rejected downtown service center and against Sidran's ordinances, now finds itself fighting publicly subsidized gentrification in South Lake Union.[41] In 2008 John Fox of the coalition complained thus that in the seven preceding years over $100 million in public city funds had gone into the redevelopment of South Lake Union, much of it supporting the property development plans of the real estate company Vulcan owned by Paul Allen. This amount almost equaled what Allen had hoped to win originally from the public when a referendum was held on his vision for an Upper East Side plus Central Park plan of a "Seattle Commons" in South Lake Union. But having rejected that plan in a referendum in 1995, the public is still providing the money and, as it were, getting the upper east side but not much park space. Going forward, Fox further fears that many more municipal dollars will be used to support Vulcan's high-rise plans for the area, with global health humanitarianism and Emerald City sustainability both playing their part in "green washing" the corporate demands for higher height limits on South Lake Union's skyscrapers.

While the old commons continue to become increasingly about public-private partnerships in South Lake Union, the boosters of the competitive global city have added the area's soaring global philanthropy reputation to their own playbook of promotional props. A local "Prosperity" blog put it like this in late 2009:

There's been a lot of buzz about the region's growing global health industry over the last two years, and for good reason: with the largest philanthropic investor in this industry in the world (Gates Foundation) here in our backyard, combined with leading research institutions in global health (University of Washington's School of Global Health, Seattle Biomedical Research Institute), and internationally recognized service delivery organizations (PATH), we're pretty well situated to become THE epicenter for this industry in the world. But a lot of people are still confused as to how a cluster that's focused on helping the world's poorest people survive the world's most widespread diseases translates into economic prosperity for our region.[42]

Confused as they might be at the start, their confusion seems unlikely to last long after they turn to the Web sites of the Greater Seattle Chamber of Commerce or the Trade and Development Alliance, where all sorts of high-ranking success stories of Seattle's global health business class are recorded, alongside the enduring hype about the region's natural future as a victor in other global city championships. In this sense, the curative global city assets merely help complement an earlier reliance on Boeing, Microsoft, Amazon, and Starbucks as growth generators. Now in addition, we are told, we can rely further on Zymogenetics, Heartstream, Alcide, Immunex, Physio-Control, and the Siemens Medical Systems' Ultrasound Group, and all these private corporations can in turn rely for expert support and labor market synergies on the public-private, philanthro-capitalist complex of South Lake Union.

In February 2010 a new Global Health Nexus venture was launched, "aimed at bringing together the expertise of local life-science companies, global-health researchers, businesses and nonprofit groups."[43] The venture further highlighted the ongoing effort to harness curative philanthropy to the larger project of selling Seattle and competing globally for foreign markets and cosmopolitan investment dollars. Summing up this promotional vision, Bob Aylward, the executive vice president for business operations for the Seattle Mariners, described the rationale for the Global Health Nexus to the *Seattle Times*: "We're sitting on the edge of a magnificent opportunity for this region to take this differentiated product, global health, and brand it as a sector that can go out and attract worldwide attention." Meanwhile Chris Rivera, president of the Washington Biotechnology and Biomedical Association, explained to the same *Seattle Times* reporter that industry really did see denationalized global opportunity in the curative global city's global reach. "For a for-profit private biotech company, the fastest emerging markets are the developing nations," he said. "We have to figure out how to partner with the organizations that already understand these markets."

From curing the world to competing in world markets, the denationalized but clearly also commodified and corporatized vision of the Global Health Nexus seems to take us back to the age-old competitive concern about promoting Seattle as a global market leader.

The co-optation of global health for selling the city in the old game of global boosterism will seem for many a bitter pill to swallow. But if the historical geography of Seattle's ongoing remaking as a global city tells us anything, it is that the definition of citizenship is always in flux, always being contested, and always, therefore, up for grabs. The Chamber of Commerce and the downtown elites may now be championing denationalized global citizenship in ways that trump global health with global competition, but the advocates of a collaborative global Seattle have not given up on the city. In 1999 the citizenry made Seattle stand for something different, and the memory of "No Globalization without Representation" continues to serve as a reminder of other more critical and collaborative conceptualizations of the global city that won't go away. Global soul is not always for sale, and, this city reminds us that there is going to be a battle for it—an ongoing Battle of Seattle over the meaning of world class.

NOTES

Many thanks to Stephen Bezruchka, Mikhail Blyth, Michael Brown, Rowan Ellis, Trevor Griffey, Judy Gunderson, Steve Herbert, Vicky Lawson, Frances McCue, Stephen Young, and Dick Walker for their ideas for improving this chapter. All arguments and errors remain my own.

1   For her classic account of global cities as command and control hubs in global hierarchies of finance, services, and new consumption practices, see Saskia Sassen, *The Global City* (Princeton, NJ: Princeton University Press, 2001). For her more recent theorization of denationalization, see Sassen, "Incompleteness and the Possibility of Making: Towards Denationalized Citizenship?" *Political Power and Social Theory* 20 (2009): 229–58. And for further geographical commentary on this theory, see Tim Cresswell, "The Prosthetic Citizen: New Geographies of Citizenship," *Political Power and Social Theory* 20 (2009): 259–73; and Matthew Sparke, "On Denationalization as Neoliberalization: Biopolitics, Class Interest, and the Incompleteness of Citizenship," *Political Power and Social Theory* 20 (2009): 287–300.

While focused on a so-called First World city, this present chapter is part of a broader project to decenter such cities as normative models in competitive global rankings. This can most obviously be done by refocusing attention on city regions outside of North America and Northern Europe (e.g., Matthew Sparke et al., "Triangulating the Borderless World: Geographies of Power in the Indonesia-Malaysia-Singapore Growth Triangle," *Transactions of the Institute of British Geographers*, n.s. 29 (2004): 485–98; and Ananya Roy, "The 21st-Century Metropolis: New Geographies of Theory," *Regional Studies* 43 (2009): 819–30. But, as Ananya Roy suggests, we also need to revisit how cities "everywhere"— including in the West—are "worlded" by social ties and political tensions that variously transect and transcend traditional global city competitiveness hierarchies. By engaging with questions of denationalized citizenship in the worlding of collaborative Seattle and curative Seattle, this chapter seeks to contribute in this cross-cutting way.

2   David Harvey, *Cosmopolitanism and the Geographies of Freedom* (New York: Columbia University Press, 2009).

3   For details, see Paul De Armond, "Netwar in the Emerald City: WTO Protest Strategy and Tactics," in John Arquilla and David Ronfeldt, *Networks and Netwars* (Santa Monica, CA: Rand Corporation, 1999). In what may well be read as another co-optation of the radicalism of 1999, this Rand corporation publication is actually a reprinted and slightly revised edition of a movement-oriented essay by De Armond, titled *Black Flag over Seattle*, still available at http://www.albion-monitor.com/seattlewto/index.html.

4   Sandy Doughton, "Global Health Stars Converge on Seattle under Cloak of Secrecy," *The Seattle Times*, June 17, 2009, http://seattletimes.nwsource.com/htm/health/2009348027_healthdavos17mo.html. See also Sally James, "Curing the World," *Washington CEO* 18, no. 1: 24–29.

5   David McCoy, Gayatri Kembhavi, Jinesh Patel, and Akish Luintel, "The Bill & Melinda Gates Foundation's Grant-Making Programme for Global Health," *The Lancet* 373 (May 9, 2009): 1645–53.

6   Suzanne Monson, "Cosmopolitan Flavor: A World Class City Offers Global Experiences for Its Downtown Residents," *Seattle Times*, New Homes Saturday Section, February 6, 2010, D4.

7   John Findlay, *Magic Lands* (Berkeley: University of California Press, 1992).

8   John M. Findlay, "The Off-Center Seattle Center," *The Pacific Northwest Quarterly* 80 (1989): 2–11.

9   See David Harvey, "From Managerialism to Entrepreneurialism," *Geografiska Annaler* B, 71 (1989): 3–10.

10   On this, see Frank Moulaert, Arantxa Rodriguez, and Erik Swyngedouw, *The Globalized City* (Oxford: Oxford University Press, 2005).

11   Timothy Gibson, *Securing the Spectacular City* (Lanham, MD: Lexington Books, 2004).

12   Rowan Ellis, "Civil Society, Savage City," PhD diss., University of Washington, 2010; Don Mitchell, *The Right to the City* (New York: Guilford Press, 2003); and Neil Smith, *The New Urban Frontier* (New York: Routledge, 1996).

13   Gibson, *Securing the Spectacular City*, 175.

14   Ibid., 232, 233.

15   Many thanks to Matt Townley for his work in calculating these data.

16   Gibson, *Securing the Spectacular City*, 178.

17   Matthew Sparke, "Not a State, but a State of Mind," in *Globalisation, Regionalisation and Cross-border Regions*, ed. Markus Perkmann and Ngai-Ling Sum (New York: Palgrave Publishers, 2002), 212–40; and Matthew Sparke, "A Neoliberal Nexus," *Political Geography* 25 (2006): 151–80.

18   Matthew Klingle, *Emerald City* (New Haven, CT: Yale University Press, 2007).

19   Trade and Development Alliance of Greater Seattle, http://www.seattletradealliance.com/aboutSea/about-greater-seattle.php.

20   Andrea James, "Seattle Named a Top 'Global Gateway' City," *Seattle Post-Intelligencer*, November 9, 2006, http://www.seattlepi.com/business/291668_officespace09.html.

21   James Lyons, "'THINK SEATTLE, ACT GLOBALLY': Speciality Coffee, Commodity Biographies and the Promotion of Place," *Cultural Studies* 19 (2005): 14–34.

22   Claire Cain Miller, "Now at Starbucks: A Rebound," *New York Times*, January 20, 2010, http://www.nytimes.com/2010/01/21/business/21sbux.html.

23   See http://www.seattlechamber.com.

24   David Solnit and Rebecca Solnit, *The Battle of the Story of The Battle of Seattle* (Oakland: AK Press, 2009).

25   Fred Moody, *Seattle and the Demons of Ambition* (New York: St. Martin's Press, 2004), 4.

26   Michael Hardt and Antonio Negri, *Multitude* (New York: Penguin Press, 2004).

27   Ibid., 287.

28   Solnit and Solnit, *The Battle of . . .* , 2.

29   See, e.g., Matthew Stadler, "Love & War," *The Stranger*, December 9–15, 1999, http://www.thestranger.com/seattle/Content?oid=2733.

30   See Trevor Griffey, "The WTO Effect," *Publicola*,

November 24, 2009, http://www.publicola.net/
2009/11/24/the-wto-effect/.

31   Solnit and Solnit, *The Battle of . . .* , 11.

32   Hardt and Negri, *Multitude,* 286.

33   Ibid., 288.

34   Thomas Friedman, "Senseless in Seattle," *New York Times,* December 1, 1999, http://www.nytimes
.com/1999/12/01/opinion/foreign-affairs-senseless-in-
seattle.html?pagewanted=1.

35   Thomas Friedman, *The World Is Flat* (New York: Farrar, Strauss and Giroux, 2005). For a critique, see Matthew Sparke, "Everywhere but Always Somewhere," *The Global South* 1 (2007): 117–26.

36   Joe Cerrell, "Making markets work, Bill & Melinda Gates Foundation," 2007, http://www.imf.org/exter-
nal/pubs/ft/fandd/2007/12/view.htm.

37   For the full report and other maps of Washington-
linked global-health research and intervention projects, see http://www.wghalliance.org/about/
where-we-work.

38   Compare with Steven Johnson, *The Ghost Map* (New York: Riverhead Books, 2006).

39   Tom Paulson, "Global health seen as big business for Seattle," *Seattle Post-Intelligencer*, October 23, 2008, http://www.seattlepi.com/local/384753_healthbiz24
.html.

40   McCoy et al., "The Bill & Melinda Gates Foundation's Grant-Making Programme," 1649.

41   See Seattle Displacement Coalition Web archive at http://zipcon.net/~jvf4119/outside_city_hall.htm.

42   Eric Schinfeld, "Global Health: Our Window to the World," November 20, 2009, http://prosperityblog
.wordpress.com/2009/11/20/global-health-window-
to-the-world.

43   These and the following quotes are from Kristi Heim, "Global Health Envisioned as City's Next Hot Industry," *Seattle Times,* February 25, 2010.

FOUR

RURAL GEOGRAPHIES

David Barker, Anne Bonds, Jennifer Devine,
Lucy Jarosz, Victoria Lawson, Lise Nelson,
and Peter Nelson

At first glance, it may seem strange to have a chapter on rural geographies in a book about Seattle. Yet one cannot understand this urban place without appreciating its broader regional context. The rural Pacific Northwest is intimately linked with city life in Seattle through economic, demographic, and political processes. Hay, wheat, and lumber, exported through the port of Seattle, contribute to the regional and urban economies. City residents who vacation, reside, and retire in rural places across the region redefine land use, rural aesthetics, and the politics of conservation. Migrants who come to the region for work on farms and in factories reshape the cultural vibrancy of the city and beyond. Without these economic and cultural connections, the city would be different.

These links are sustained in part by the dominant images of towering mountains, salmon streams, and wilderness trails characterizing the Pacific Northwest. But these cultural icons obscure other parts of the rural landscape. The rural regions of the Northwest, encompassing Oregon, Washington, Idaho, and Montana, are also places of impoverishment, low wages, hardship, and economic transformation. The contemporary landscapes of the American Northwest are being reworked in a range of ways, shifting from largely farm work and the extraction of resources to tourism and service work. These shifts are driven by the exhaustion of finite resources, changing environmental attitudes and pressures, changing farm subsidies, a reduced availability of public lands for grazing, and reduced access to public lands for logging. These changes bring the Old West of miners, cowboys, timber workers, farmers, and ranchers into contact (and sometimes tension) with the New West of tourists, vacation homes, ski resorts, and upscale recreational amenities, retailers, and theme towns. Here we highlight these often-hidden rural geographies by portraying them in three types of landscapes: playgrounds, dumping grounds, and unseen grounds.

Recent Demographic and Economic Trends in the Rural Northwest

Stretching from the Pacific Coast of Oregon and Washington through the glacier-clad Cascade Mountains and on into the basin and range landscapes of Idaho, Montana, and Wyoming, this five-state region possesses perhaps the most diverse physical geography in the entire United States. Like much of the rural western United States, the nonmetropolitan portions of these five northwestern states have experienced significant demographic and economic changes over the past two decades, which can be summarized as follows:

1. The rural Northwest saw robust population growth between 1990 and 2000. This growth is most pronounced in high-amenity destinations in close proximity to public lands. The growth subsided considerably between 2000 and 2008, but continued in high-amenity areas.
2. Hispanics are becoming an increasing presence in the region and are beginning to spread into areas beyond the more agriculturally dependent communities.
3. A few pockets of extreme wealth are emerging in the region's resort destinations, suggesting that class polarization is spreading out across space.

POPULATION CHANGE IN THE RURAL NORTHWEST
Overall, the last two decades brought robust population growth to the rural Northwest. Mirroring national patterns of a rural rebound in the 1990s, the region's population expanded 14.1 percent during that decade, and this growth carried over into the early part of the twenty-first century at a 6.8 percent rate. These regional growth rates greatly outpaced the national growth experiences over the same time. Nearly 60 percent of the nonmetropolitan counties

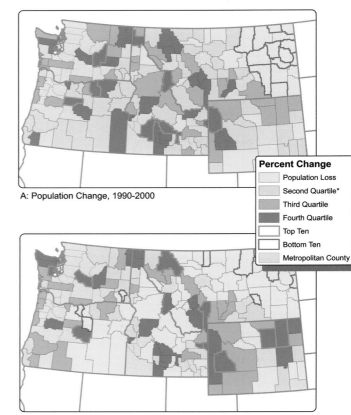

A: Population Change, 1990-2000

Percent Change
Population Loss
Second Quartile*
Third Quartile
Fourth Quartile
Top Ten
Bottom Ten
Metropolitan County

B: Population Change, 2000-2008

4.1 Percent change in total population for the nonmetropolitan Northwest. All counties shown in green had positive population growth; the darker the shade of green, the more rapid the population growth.
NOTE: The data cut-off between the first and second quartiles was modified to allow differentiation between population loss (shown in yellow) and population gain.

in the Northwest had growth rates exceeding the national nonmetropolitan growth rate of 7.5 percent during the 1990s, and only 17 percent actually lost population. Though still surpassing the national growth trends, since 2000 nonmetropolitan growth in the Northwest has subsided considerably. In this more recent period, only 43 percent of the region's counties had growth rates exceeding the national nonmetropolitan rate of 2.5 percent and roughly the same number of counties—42 percent—in the region lost population.

TABLE 4.1 POPULATION CHANGE FOR THE FASTEST AND SLOWEST GROWING COUNTIES IN THE NONMETROPOLITAN NORTHWEST, 1990-2000 VS. 2000-2008

1990-2000				2000-2008			
County	State	% Change	Domestic Migration Rate (%)	County	State	% Change	Domestic Migration Rate (%)
Top 10				**Top 10**			
Teton	ID	73.48	42.60	Teton	ID	47.24	26.45
Teton	WY	61.99	18.51	Sublette	WY	42.84	37.85
Morrow	OR	44.42	24.47	Madison	ID	36.37	7.64
Ravalli	MT	43.81	40.92	Gallatin	MT	32.42	23.88
San Juan	WA	38.65	24.64	Campbell	WY	23.07	12.76
Blaine	ID	37.97	14.08	Crook	OR	20.01	17.54
Bonner	ID	37.66	31.13	Johnson	WY	19.63	19.75
Jefferson	OR	37.59	8.11	Flathead	MT	18.80	14.87
Elmore	ID	37.23	-1.70	Mason	WA	17.09	15.69
Grant	WA	35.74	12.50	Kittitas	WA	16.75	12.65
Bottom 10				**Bottom 10**			
Richland	MT	-9.21	-7.31	Garfield	WA	-14.06	-10.22
Daniels	MT	-10.12	-6.42	Wheeler	OR	-14.74	-8.27
Wibaux	MT	-10.25	-3.19	Phillips	MT	-15.75	-12.91
Powder River	MT	-10.41	-14.56	McCone	MT	-15.23	-14.16
Rosebud	MT	-10.53	-14.74	Sherman	OR	-15.31	-15.62
Phillips	MT	-10.68	-10.81	Daniels	MT	-18.54	-13.24
McCone	MT	-12.37	-14.89	Wibaux	MT	-18.91	-11.14
Prairie	MT	-12.48	2.92	Sheridan	MT	-20.02	-12.81
Sheridan	MT	-12.51	-8.14	Liberty	MT	-20.06	-17.42
Garfield	MT	-19.41	-12.22	Treasure	MT	-26.02	-22.88

Population change shows distinct geographic variation across the rural Northwest (see fig. 4.1 and table 4.1). The highest growth rates are found along the spine of the northern Rockies in Wyoming, Idaho, and Montana, as well as along the Pacific Coast. In contrast, the declining counties are heavily concentrated in the far eastern stretches of the region.

A combination of factors can explain the geographic variation in growth experiences across the rural Northwest. First, proximity and access to urban centers greatly influence the patterns of population growth and decline. Eight of the ten fastest growing counties in the nonmetropolitan Northwest in the period 2000 to 2008 contain "micropolitan" centers—small cities that act as a regional urban hub.

In contrast, five of the ten counties with the highest population loss over the same period were classified as the least urban by the U.S. Department of Agriculture.

These counties have no population center greater than 2,500 people and are not adjacent to a metropolitan or micropolitan area. Second, population growth driven by domestic migration also appears to be drawn toward high amenity areas. In the Northwest, natural amenities often include large tracts of public lands, national parks and monuments, wilderness areas, scenic vistas, and other features. This phenomenon is perhaps best illustrated in the greater Yellowstone region straddling the borders between Idaho, Montana, and Wyoming. Teton County, Wyoming; Teton County, Idaho; and Gallatin County, Montana, have experienced extremely high domestic migration as households seek out these high-amenity landscapes. Third, many of these areas with reasonable access to medium or larger urban centers have become premium retirement destinations. With nearly 80 million baby boomers aging into retirement in the second decade of the twenty-first century, these areas are primed for significant growth, and it is not surprising that four of the fastest growing counties in the region since 2000 are classified as retirement dependent. Each of these four is either micropolitan or adjacent to a micropolitan area demonstrating the confluence of these forces driving population change in the region.

INCREASING ETHNIC DIVERSITY
IN THE RURAL NORTHWEST

The patterns of growth across the rural Northwest have been layered on top of increasing ethnic diversity, driven in large part by a growing Latino population in many communities. For several decades, Latinos have found work in the region's agriculture and food-processing industries, and this is evidenced in panel A of figure 4.2. and table 4.2. Many of the counties with the highest concentrations of Latinos are classified as farm-dependent counties.[1] Virtually all of the top ten counties in figure 4.2 are found in the fertile lands of Oregon and Washington, along the Columbia River, or along southern Idaho's Snake River. These

regions produce large quantities of such labor-intensive crops as grapes, apples, peaches, cherries, potatoes, onions, and a variety of berries, all of which need to be picked by hand.

While the highest concentrations of Latinos are found in the more agriculturally dependent regions, it appears that Latinos are spreading out into other areas of the rural Northwest. Panel 2 of table 4.2 and panel B of figure 4.2 identify the areas with the most rapid growth of Latinos between 2000 and 2008. Panel B shows the Latino popula-

A: Percent of Total Population Latino, 2008

Percent
- First Quartile
- Second Quartile
- Third Quartile
- Fourth Quartile
- Top Ten
- Metropolitan County

B: Percent Change Latino, 2000-2008

4.2 Latino population concentrations and relative change in the nonmetropolitan Northwest. The darker the shading, the higher the concentration of Latinos (panel A) or the more rapid the growth of the Latino population (panel B).

TABLE 4.2 CONCENTRATION AND CHANGE IN THE LATINO POPULATION, 1990-2008

2000			2008		
PERCENT OF TOTAL POPULATION					
County	State	Percent	County	State	Percent
Top 10					
Adams*	WA	47.08	Adams*	WA	55.12
Clark*	ID	34.25	Clark*	ID	40.44
Grant*	WA	30.09	Grant*	WA	35.69
Malheur	OR	25.62	Minidoka*	ID	30.25
Minidoka*	ID	25.46	Morrow*	OR	29.77
Hood River	OR	25.02	Malheur	OR	27.76
Morrow*	OR	24.43	Jerome*	ID	27.31
Cassia*	ID	18.74	Hood River	OR	26.72
Jefferson	OR	17.74	Gooding*	ID	24.27
Jerome*	ID	17.17	Cassia*	ID	23.39
PERCENT CHANGE, LATINOS					
Camas*†	ID	1275.00	Sublette†	WY	239.29
Sweet Grass*†	MT	671.43	Glacier†	MT	198.11
Wheeler*	OR	558.33	Valley†	ID	145.33
Teton†	WY	547.54	Judith Basin*	MT	130.77
Clark*†	ID	514.04	Flathead†	MT	122.53
Blaine†	ID	410.05	Gallatin†	MT	120.82
Granite*†	MT	300.00	Teton†	WY	114.35
Lincoln*	OR	255.54	Adams†	ID	111.11
Sherman†	OR	235.71	Lincoln	WY	102.54
Tillamook†	OR	229.10	Roosevelt	MT	100.76

* Denotes farm-dependent counties

† Denotes recreation-dependent counties

tion growing most rapidly in many of the high-amenity areas. In these new destinations, Latinos are presumably responding to increasing employment opportunities in the service industries and construction trades stimulated by the growth of the overall population. Of the top ten counties ranked in terms of Latino population change, only one is classified as farm dependent and seven as recreation dependent counties (in 2008). Most of these recreation counties are found in the mountains and along the coasts, a long way from the region's agricultural centers. Teton County, Wyoming, provides a striking illustration of the rapid pace of Latino population change. In 1990 the decennial Census recorded just 183 Latinos in Teton County out of a population of 11,267 (1.6 percent). By 2000 the Census showed 1,185 Latinos living in the county, and the most recent estimates available for 2008 indicate a Latino population of 2,540 out of 20,376 total residents, or 12.5 percent. Clearly, the expanding Latino presence is beginning to impact new destinations.

EMERGING ISLANDS OF AFFLUENCE

In the Northwest, new activities are emerging as important contributors to rural economic bases. Since the 1960s, employment in manufacturing and service industries has grown considerably in rural areas across the country, while traditional resource-based sectors have declined in relative and absolute importance. In addition, nonemployment income in the form of dividends, interests, and rents has become an increasingly important component of nonmetropolitan income streams, and migration is drawing more nonemployment income into rural areas. As affluent individuals seek out attractive places to live, they often bring with them accumulated wealth, producing highly visible pockets of affluence in specific rural locales. Some of the most pronounced "islands of affluence" are found in the rural Northwest and are presented in figure 4.3 and table 4.3.

In the year 2007, roughly 17.5 percent of personal income in the United States was derived from dividends, interests, and rent. In places like Sun Valley, Idaho (Blaine County), or Jackson Hole, Wyoming (Teton County), 50 percent or more of annual personal income is derived from accumulated wealth. Assuming the $1.7 billion in dividends, interest, and rent arriving in Teton County is distributed equally among its residents, the per capita level of dividends, interest, and rent in the county exceeds $85,000 per year. Many of the areas with high dependence on nonemployment income coincide with the amenity-rich, recreation-dependent destinations that have accelerated population growth and growing Latino populations. In fact, eight of the top ten counties listed in table 3 for the year 2007 are recreation dependent, and all counties in the top ten have more than twice the national share of income from dividends, interest, and rent. Given that the origin of dividends, interest, and rent income is likely to be nonlocal, these income sources represent new money flowing into the county economies, or new sources of economic base. Furthermore, given the likelihood that this income is *not* evenly distributed across the population in places like Teton County, such communities clearly include some

4.3 Percent of total personal income derived from dividends, interest, and rent, 2007. The darker the shading, the greater the share of total personal income derived from dividends, interest, and rent.

Share of Total Personal Income
First Quartile
Second Quartile
Third Quartile
Fourth Quartile
Top 10
Metropolitan County

TABLE 4.3 COUNTIES WITH HIGHEST DEPENDENCE ON NONEMPLOYMENT INCOME

2000				2007		
Share of Personal Income (%)				**Share of Personal Income (%)**		
Top 10				**Top 10**		
Teton	WY	51.09		Teton	WY	64.30
San Juan	WA	48.51		San Juan	WA	54.67
Sweet Grass	MT	40.45		Blaine	ID	45.47
Gilliam	OR	40.03		Sweet Grass	MT	38.77
Wheeler	OR	39.38		Sheridan	WY	38.61
Golden Valley	MT	37.03		Golden Valley	MT	37.49
Carter	MT	36.95		Park	WY	36.66
Johnson	WY	36.84		Jefferson	WA	36.05
Garfield	MT	36.64		Johnson	WY	35.56
Judith Basin	MT	36.05		Wheatland	MT	35.53

extremely wealthy people, further contributing to the class polarization in the region.

It is also noteworthy that table 4.3 shows a few counties in the eastern portion of the region with high dependence on dividends, interest, and rent. These places should not be confused with the likes of Jackson Hole or Sun Valley. For the counties in the eastern portion of the region, the high dependence on dividends, interest, and rents indicate broader forces of economic restructuring in the resource-based industries. The rental income generated in places like Golden Valley and Wheatland County, Montana, is likely derived from the rental of farmland to increasingly large farm operators. Between 2002 and 2007, the number of farmland owners in Golden Valley, Montana, receiving rental income from their land more than doubled, as smaller operators ceased farming their own property and leased it to larger operations. Such processes are symptomatic of agricultural restructuring and lie behind some of the other demographic trends described above.

LAYERING THE OLD AND THE NEW
Some have described all these demographic and economic

changes as a shift from "Old West" to "New West." The Old West is often characterized by resource extraction activities, as well as by heavy federal government expenditures on dams and other facilities, while the New West is seen as dependent on a service-based economy, attracting large numbers of footloose and affluent individuals to the high-amenity landscapes in the region. These two labels suggest that the West is a place of stark contrasts and oppositions.

Such a clear distinction between old and new neglects the considerable practical fuzziness in the region. Counties like Teton, Wyoming, and Blaine, Idaho, have become places of exceptional wealth, and the shift from resource extraction to natural amenities is almost complete. It would be premature, however, to pen the obituary for the "Old" West. Traveling through the Teton counties, or from Bend, Oregon, up to the Cascade crest, the landscape remains dominated by heavily resource-intensive activities. While these activities may not be driving the growth in the region, they are still visibly dominant forces shaping the landscape. Furthermore, transformations in these resource-dependent activities are a powerful catalyst behind specific demographic changes, including the increasing ethnic diversity in certain areas and the population decline in eastern portions of the region. Therefore, a new West does not replace an old West. Rather, new economic and demographic forces work with the more traditional forces that originally developed the region. These processes of change require careful analysis and defy simple binary classifications of "old" and "new."

Beyond the Dominant Image: Theorizing Rural Difference

There are at least three distinctive landscapes in the rural places of the Northwest that are connected to economic restructuring, rural poverty, and contemporary economic development: playgrounds, dumping grounds, and unseen grounds. This typology of the rural American West is not exhaustive; rather, it is indicative of the diversity of realities and experiences in rural places. The rural Pacific Northwest is a destination point and home to a diversity of people of varying economic positions, races, and ethnicities in places that are alluring and beautiful and places that are invisible and unseen. We explore the Northwest landscape through the prism of these types of places, in order to show how the popular images and the iconic landscapes obscure as much as they reveal. If the common image is the playground landscape, opening our eyes to both unseen grounds and dumping grounds reveals categories of race and poverty that are integrally linked to middle class, predominately white lifestyles of consumption. In what follows, we provide some examples of the broader processes that are making geographies of poverty in the region.

PLAYGROUNDS
Playgrounds are places such as Gallatin and Flathead counties, Montana; the Methow Valley; San Juan Islands and Suncadia in Washington; Bend, Oregon; Driggs and Coeur D'Alene, Idaho. They are aesthetically appealing areas of costly leisure activities and multimillion-dollar homes and ranches. Playgrounds are located near mountains, lakes, coastal shores, forests, national parks, and ski resorts (fig. 4.4). They are easily recognizable for their beauty and wildness. Playgrounds are dominant in the public imaginaries of the Northwest. People want to live in these places because of their leisure-time activities and aesthetics. These areas are experiencing substantial gentrification, as they become the sites of second, third, and even fourth homes of the wealthy and famous.[2]

Tourism is an important aspect of any economy, and property values are significantly higher in playgrounds than in other rural areas. Playgrounds may also be locations for businesses primarily serving wealthy or middle-class consumers, such as art galleries, boutique clothing and sporting goods stores, gourmet restaurants, and artisanal bakeries. Environmental awareness, liberal political values, middle-class lifestyles emphasizing outdoor activities, and the support of local business and small-scale, local and sustainable food production define these areas.

Kittitas County in Washington State, for example, has gone from being largely ranchland to being a playground in three decades. This county is located on the eastern slopes of the majestic Cascade Mountain range. Its diverse topography spans vast timberlands covering half of the county's 2,315 miles and prime pastures of bunchgrass ideal for grazing livestock and growing hay. Surpassing earlier boom periods in logging and mining, cattle ranching and hay and grain production drove the county's economy until the 1970s. Since then, Kittitas's natural amenities, proximity to Puget Sound, and celebrated ranching history have enabled developers to transform the ranchland into a spot sought after by tourists, bedroom commuters, hunters, second-homers, and retirees. This economic transformation is intimately connected with economic restructuring in the U.S. economy away from resource extraction, manufacturing, and farming to diverse service-based sectors. As Seattle became a global hub of computer software development and biotechnology research and the headquarters for Star-

4.4 View from Mount Rainier.

bucks and REI, Kittitas County's location as a recreation hotspot on the map of America's "New West" has turned on the successful marketing and creative revitalization of the county's "Old West" ranching history and cowboy culture.

DUMPING GROUNDS

Dumping grounds are the places constituting the working landscapes of the West, places like Stevens and Benton counties, Washington; Cassia County, Idaho; and Big Horn County, Montana. These places are former farming areas, ranchlands and/or resource-extraction areas that offer labor, land, and water resources in order to attract prisons, farming factories, and food processing plants and thus

bring economic development to rural areas of high unemployment and poverty. Economies of dumping grounds are moving increasingly toward retail and service-sector activities dominated by discount stores such as Wal-Mart and fast food chains such as McDonald's and Burger King. The areas may be near beautiful mountains or rivers, but they lack the spectacular scenery of playgrounds and proximity to large urban areas or well-known national parks. Rural economic restructuring in these areas is signaled by the closure of mines and a downturn in logging activities, leaving rural places starved for jobs and a secure tax base. City and county leaders compete with other rural areas around the country to attract national and transnational agribusiness and manufacturing firms, state and federal prisons, and other forms of business through the lure of cheap land, water, and labor, as well as by offering tax breaks or infra-

4.5 Prison in Hardin, Montana.

structural development (such as roads). These places often welcome businesses that those living in urban areas and in playgrounds see as polluting and undesirable: large-scale hog farms, food processing plants, and solid-waste sites—hence the epithet: dumping ground. See the following case study for links between prison investments and ongoing rural poverty (see fig. 4.5).

Over the past three decades, prison populations in the United States have soared to unprecedented levels. In fact, between 1970 and 2003, the U.S. prison population grew sevenfold, making it the largest in the world.[3] Driven by "tough on crime" laws that sent more people to prison and kept them there longer, local, state, and federal governments have scrambled to keep pace with expanding levels of incarceration and increasingly overcrowded prisons, thereby facilitating a prison construction boom. Though this boom peaked in the 1990s, the monumental expansion of prisons continues to reshape landscapes across the United States and now affects extended families and encompasses entire neighborhoods and communities. It also raises a series of pressing questions about where to locate prisons, how to fund prison building, and what the human costs of this growth are.

Prison construction has boomed particularly in rural communities. An average of 25 new rural prisons opened *each year* in the 1990s, leading to the siting of 350 new rural prisons.[4] During the 1990s, a new prison opened within rural America every fifteen days.[5] Why are so many prisons located in rural America? Many rural communities are negotiating the economic changes outlined above, as they also confront severely limited budgets, declining tax revenue, a less diversified economy, and restricted employment opportunities.

As certain rural economies languish and prison construction continues to expand, prisons have emerged as a "recession-proof" opportunity for communities seeking industrial investment and job creation and have become an increasingly common component of economic growth initiatives. In fact, some rural communities actively recruit prisons, utilizing tax breaks, land, infrastructural guarantees, and a series of other incentives to secure public and private prison bids.

These trends are playing out in important ways in Washington, Oregon, and Montana as each state has experienced substantial prison development in rural communities. For example, in Ridge County, Montana, town leaders in Carson worked hard to recruit a prison as the solution to a depressed local economy, high unemployment, and deep poverty. Ultimately, rather than creating local economic growth, their efforts have greatly damaged the sense of community and created more questions about the area's economic future.

In Montana, the already volatile nature of agriculture and ranching has been compounded by increased flows of wheat and beef from Canada under the North American Free Trade Agreement (NAFTA) that have driven down prices for local ranchers.[6] Additionally, there was a 24 percent decrease in the state's mining firms between 1992 and 2002 (U.S. Bureau of the Census). Ridge County, in the eastern part of the state, is experiencing these restructurings acutely.

Ridge County's history is shaped by the colonization of its indigenous peoples, struggles over land and resources, racial inequality, and persistent poverty. Nearly two-thirds of the county is designated as tribal land. Despite its small population of just over 3,000, inequality is conspicuous. Like other reservation counties, the area has some of the most historically entrenched levels of poverty in the state and, indeed, in the nation. The county's 2005 poverty rate stood at 29 percent, more than double the national poverty rate of 12.4 percent, or the state rate of 14 percent.[7]

These economic forces—a depressed local economy and deeply entrenched poverty—set the stage for a proposed prison in the town of Carson. Faced with declines in primary industries and few amenities to compete with more scenic, gentrifying areas in the state, town leaders in Carson try to recruit any kind of investment they can. Rural communities like Carson have become dumping grounds as a consequence. Community leaders work to loosen environmental, zoning, and labor restrictions in order to attract any industrial development at all, thus bringing in prisons, corporate dairies and hog farms, toxic waste storage, and food processing and animal slaughtering plants. This results in a race to the bottom, wherein leaders pursue competitive, pro-growth economic development projects at any cost, resulting in dramatic economic shifts, undesirable employment opportunities, and fewer living wage jobs. Ironically, leaders build support for these projects by emphasizing job creation and employment growth.

Ridge County has been the potential site for two private corrections facilities in the past decade. Just a decade earlier, the community was nearly selected as the location for a new state prison to be managed by the Corrections Corporation of America (CCA). However, county commissioners, representing their largely tribal constituency, prohibited the prison from moving forward on county land just outside of Carson. More recently, local officials were approached by yet another private corrections corporation with a feasibility study highlighting the appropriateness of Carson for a correctional facility. Ultimately, town leaders made further efforts to build a 464-bed detention center, despite having no secured inmate contracts with the state of Montana. The proposed institution was to bring 110 new jobs to the area, and research participants repeatedly said that the prison was "strictly an economic development project."[8]

With construction complete and employees in place, the facility was scheduled to open in August of 2007. However, the detention center continues to sit empty as Carson struggles to obtain inmate contracts, with a $27 million bond that went into default shortly after the facility was to be opened in 2007. Although the community and the private firm had tentative assurance that the Department of Corrections would need the space, this agreement collapsed after the election of a new governor in 2005. In efforts to address the lack of inmates, the private correctional company has been seeking out-of-state contracts. However, later in 2007, the Montana State attorney general ruled that the facility cannot legally house out-of-state inmates. Since then, town officials have made offers to host ever more politically marginalized populations, such as high-level sex offenders and including the 240 detainees from the Guantánamo Bay Detention Center.

As Carson's local officials struggle with the disastrous outcome of their prison development efforts, the barbed wire–surrounded facility is empty, literally in the middle of town. Meanwhile, Ridge County's poverty rate continues to soar, unemployment rates climb, and service providers struggle to offer much needed provisions to the community's growing numbers of people in need. Carson, now burdened with a debt that it cannot pay, continues to seek inmates through a series of increasingly desperate efforts. Sadly, those suffering most from the prison development efforts are the poorest community members.

MIGRANT WORKERS IN DUMPING GROUNDS

One of the most paradigmatic examples of rural inequality is the dependence of modern agriculture, food processing, and tree-planting industries on low-wage, immigrant (usually Latino) labor. Dependence on Latino farmworkers in the Pacific Northwest reaches back to the 1942 Bracero Program, which brought 47,000 Mexican workers to the region to harvest crops that threatened to spoil because of World War II labor shortages. Historically, most of these workers migrated seasonally and were largely invisible to predominantly white, rural communities. This invisibility stemmed not only from the temporary nature of their presence but from the spatial containment of farmworkers in isolated labor camps. Starting in the early 1980s, however, a growing number of these workers began to settle perma-

4.7 Downtown Woodburn, 2004.

nently in the region. By the 1990s and beyond, the figure of the male and migrant (seasonal) farmworker was largely (although not completely) replaced by farmworker *families* who were settling in small and medium-sized rural towns and communities throughout the Northwest. These resident farmworkers—a broad term used to refer to workers on farms, in food processing facilities, and in tree-planting operations—transformed the cultural, social, and political landscapes of the rural Northwest in ways often predicated on their continued socioeconomic marginalization and exclusion from dominant understandings of place and belonging (see fig. 4.6).

The influx of Latinos and struggles over housing and services in Woodburn, Oregon, a small city located at the heart of the agriculturally rich northern Willamette Valley (fig. 4.7), provides another example of a dumping ground landscape, one which illustrates tensions around race, poverty, and belonging. Around Woodburn, the demand for farmworkers expanded dramatically in the 1970s and into the 1980s as new labor-intensive industries emerged (e.g., greenhouse and nursery crops) or old industries expanded (canning, food processing, tree planting). A growing range of rural employers began actively to recruit Mexican workers because immigrant workers—particularly undocumented ones—were less expensive and less able to contest working conditions.

4.6 Map of northwestern Oregon showing northern Willamette Valley, Woodburn, and Portland. The top-five farmworker-receiving counties in the state are highlighted.

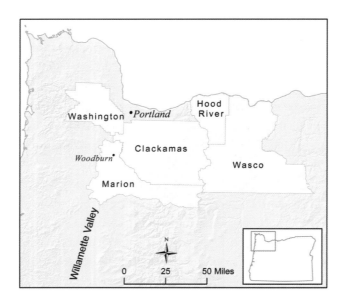

4.8 Migrant farmworker housing, ca. 1965–68.

The movement of farmworkers into Woodburn and other nearby towns overwhelmed the existing housing stock. Landlords took advantage of the situation and often charged rent on a per-room basis, crowding several families into single-family housing. Farmworker families were paying several hundred dollars to live in a garage, for example. The local newspaper, the *Woodburn Independent*, ran numerous stories in the late 1980s on the overcrowded and often dangerous housing conditions in many neighborhoods. In 1990 the city government organized a "Livability Forum," which according to the newspaper was inundated by a crowd of "angry citizens."

In response to this crisis, the Farmworker Housing Development Corporation (FHDC) was founded in 1991 by activists who were involved in a range of farmworker advocacy organizations. These activists realized that a crucial issue in their broader goal to protect the rights and health of farmworkers was access to decent, safe, and affordable housing, located—they were very explicit on this point—*within city limits*. They were determined to claim a space of belonging for farmworkers in the community and to under-

mine the labor camp as the assumed place "out there" for farmworkers (fig. 4.8). Eventually they set their sights on a city-owned piece of property within walking distance of downtown Woodburn, on which to build the Nuevo Amanecer (New Dawn) complex.

However, the city council, the mayor, and a group of vocal citizens fought for over a year to prevent FHDC from buying the property. The city relented only after the state of Oregon threatened to revoke its eligibility for additional community development block grants. Although race and the migration status of farmworkers were never explicitly mentioned during these hearings, analysis of transcripts and interviews with key actors suggested that race, "illegality," and class haunted these debates that were framed as being about "taxpayer rights." Such issues emerged more explicitly a year later when anonymous hate mail was sent to an FHDC neighbor describing Nuevo Amanecer in this way:

The Mexicans are going to have a housing project in Woodburn, right across the street from the high school where their gangs can freely mingle with our kids. . . . The

Mexicans will work the summer season and then spend the winters in living quarters built for them with our money.[9]

State-sanctioned and subsidized farmworker housing within Woodburn's city limits legitimized the presence of farmworkers in town, something that was an affront to some of the town's white residents even if the city's budget would benefit from the transaction and it would alleviate a very visible housing crisis.

Two years later, FHDC purchased a vacant lot across from the Woodburn City Hall, in the heart of downtown, in order to build a new complex, Esperanza (Hope) Court. During the intervening years, Nuevo Amanecer was an exemplary housing complex, including national recognition for its design and participatory governance structure. Laudatory articles in the *Woodburn Independent* seemed to suggest that the fears expressed by some had been unfounded: Nuevo Amanecer was a big success for the residents and larger community alike. Yet once again, FHDC experienced significant opposition to their new project on the part of the planning commission, city council, and some residents.

The struggle over Esperanza Court came to a head during a city council hearing in October 1996. Supporters, including farmworkers, activists, advocates, and health-care workers, highlighted the difficulty workers had finding decent housing. While opponents acknowledged the need for housing, they argued that *this* was not the appropriate location for farmworker housing. Although the speakers did not articulate *where* they thought farmworker families and housing belonged, perhaps they were harkening back to a time when the social spaces of farmworkers were more fully contained within labor camps—beyond city boundaries and out of sight. The impassioned speeches by opponents about the future of Woodburn and its property values were made in vain, as the city attorney advised that any action other than approval would be overturned in court. The council voted to approve the plan.

The struggles over Nuevo Amanecer and Esperanza Court brought out competing visions of place, race, and belonging in Woodburn during a period of rapid demographic change. The effort to build safe and affordable housing for farmworkers in Woodburn sheds light on ongoing and future struggles over race and belonging in the region.

UNSEEN GROUNDS

Unseen grounds are the forgotten, invisible, rural places of the Northwest. Examples are Hooper and Warren counties in Montana; areas in northeastern Washington State; and parts of southern Idaho. They are marked by shuttered shops, dying small towns, and out-migration in the wake of the decline in natural resource extraction through logging and mining. These places may face environmental problems of declining water quality or soil fertility as a result of historic forms of resource extraction. They are the marginalized environments of struggling small ranches and low-paid service sector employment, offering little economic opportunity for young families for a sustained quality of life.

Unseen grounds also possess unconventional forms of beauty, but these are effectively invisible because of economic and social marginalization. These spaces can be the arid landscapes of sagebrush and antelope, of rolling hills unfolding beneath brilliant blue skies that stretch endlessly overhead. But unseen grounds are not destination points for tourists; they are the bypassed spaces in the New West. Unseen grounds have negative population growth: they

4.9 Main Street in "unseen grounds," eastern Montana.

are disappearing communities (see table 4.1). In the Pacific Northwest we find many "unseen" counties concentrated in rural eastern Montana, where fragile ecologies and disappearing investments are expressed by stores lining Main Streets that are closed and shuttered (fig. 4.9). Simulated pearl necklaces and birthday cards still on display in the window of one general store are covered with dust, cobwebs, and grit.

We call these places "unseen grounds" because they are off the grid. They are far away from and inaccessible to major transport routes and established recreation resorts, and they are sparsely populated and politically marginalized. In addition, these places remain reliant on politically unpopular mining and logging activities. These are vulnerable environments, with harsh seasons, that have been historically damaged by efforts at economic development that led to pollution of groundwater and intensive agriculture that eroded unirrigated and fragile lands. These areas, where local economies remain heavily dependent upon ranching and small farms, have the highest rates of poverty (fig. 4.10).

Residents frequent the local bars and cafes but travel to larger towns and Wal-Mart centers for groceries, medicines, and jobs. Individual efforts to revive community businesses have failed, and small grocery stores and small cafes off the major highways display For Sale signs. Individual businessmen and artisans employ the Internet to

4.10 Dryland farming in southern Idaho.

advertise and sell their products around the country. There may not be adequate, potable water available despite large amounts of cheap land and willing, cheap labor. Highways encourage residents to drive elsewhere to shop and pursue recreational activities. Leaders are dispirited, unengaged, or lacking in vision. Community activists work toward visions of a different future but are cynical about resident apathy and disengagement.

Residents acknowledge that most people in their communities are poor. Employment opportunities are extremely constrained here. Most residents commute to work in larger, nearby towns. The best that community leaders hope for is that economic development and growth occur nearby and that their communities will develop into bedroom communities serving larger areas with cheap labor and cheaper housing. A social service provider remarks that "the poor are us." This solidarity with the very poor stood out in our research, because in other areas with high Latino poverty, whites whom we interviewed focused on differences rather than alliances.

Rethinking the Rural Northwest

Wealth and poverty, work and leisure, and the beauty of the valleys, mountains, lakes, and coastlines of the American West are woven through three distinctive landscapes that contain economic development trajectories fueled by neoliberal economic policy and that racialize rural poverty in different ways. The New West is indeed booming, as witnessed by rapid urbanization and growth in rural areas and playgrounds, but it also contains spaces of deep vulnerability and want in areas we call unseen grounds and dumping grounds. Our typology of rural places in the Pacific Northwest suggests that rural impoverishment is the recent result of economic restructuring and the movement away from historical forms of rural employment and development to new forms exemplified by playgrounds and wealthy and middle-class landscape consumption emanating from Seattle. Dumping grounds hew the neoliberal line of development and attempt to woo any investments in manufacturing or agriculture to their regions, but this time it is the forms of business that no one else wants in their backyards. Unseen grounds are areas that are transforming and disappearing from both iconic and economic development images.

Poverty unfolds unevenly in all of these places and is most visible in places that travelers to the New West will never see. But even though poverty may be denied and invisible in rural spaces, it is indeed substantially present in the rural New West. Dumping grounds and unseen grounds exist on the peripheries of our collective visions of the West. They reveal that rural poverty is not just present in the Mississippi Delta, Appalachia, or Native American reservations of the Southwest, but that it also occurs in regions that are celebrated as economic development success stories. Certainly the greater metropolitan region of Seattle is seen as this kind of success story, as are its regional and rural playgrounds and sites of agrotourism, such as the nationally known ski resorts of Idaho and Montana and the vineyards of Washington. It is important to acknowledge that sociospatial trajectories of urbanization, rural restructuring, and the movements of money, agricultural productions, and farm and food workers between the city and the countryside bind Seattle and rural landscapes together so that poverty and wealth and the city and the country remain always linked, both seen and unseen.

NOTES

1   Farm-dependent counties are identified by the Department of Agriculture as those in which 15 percent of income is derived from farm activities, or at least 15 percent of employment is in agriculture.

2   See http://www.destination360.com/north-america/us/washington/methow-valley-wa (accessed September 28, 2009).

3   Bruce Western, *Punishment and Inequality in America* (New York: The Russell Sage Foundation, 2006).

4   C. Beale, "Cellular Rural Development: New Prisons in Rural and Small Town Areas in the 1990s." Paper presented at the annual meeting of the Rural Sociology Society, Albuquerque, NM, August 18, 2001.

5   Ibid.

6   S. Gallaher, "Farmers Warned to Be Aware of Stress and Economic Turmoil." The Associated Press state and local wire, Helena, Montana. December 18, 1998. LexisNexisAcademic.

7   U.S. Bureau of the Census, Economic Census, 1992, 1997, 2002. http://www.census.gov/econ/census02.

8   Personal communication, 2006.

9   G. Rede, "Woodburn Police Investigate Anti-Hispanic Letter," *The Oregonian*, March 19, 1993, 5.

FIVE

POLITICAL GEOGRAPHIES

Richard Morrill, Larry Knopp,
Steve Herbert, John Carr, Tim Nyerges,
Kevin Ramsey, Matthew Wilson,
and Sarah Elwood

Geography is always political. The control over territory and attempts to shape its borders or what goes on within it are key issues in political geography. This field is interested in the geographical manifestations of power in all its forms, governmental, economic, and even social and cultural. Geography can be the source of power, as in electoral support for a politician, or a consequence of power, as in the exercise of control over a neighborhood. In this chapter, we examine the political geographies of Seattle in a wide variety of manifestations: the jurisdictional organization of the region, its electoral geographies, the political culture of the city, planning and transportation issues, and the role of nonprofits in urban politics.

A WEB OF JURISDICTIONS

Larry Knopp and Richard Morrill

The web of local government in Greater Seattle is highly complex, and the degree of fragmentation exceptionally great. Power is dispersed and rather constrained by the deliberate constitutional limitations on jurisdictions. The relative ease of launching both referenda and initiatives also contributes to the dispersion of power. Unlike much of the Northeast, with towns, townships, and boroughs, Washington has no system of general purpose governments that exhaustively cover the state, and counties have historically been quite limited in their powers. Rather, the state has created large numbers of limited-purpose entities, such as "special districts" and interjurisdictional entities, instead of granting more comprehensive power to existing governments. The most commonly offered rationale for this decentralized system is a quasi-libertarian fear of large government, even though voters seem rather willing to tax themselves significantly. Table 5.1 lists the kinds and numbers of governments in Greater Seattle.

Federal government influence is vast everywhere in the country, but is especially evident in much of Greater Seattle through the actual control over territory. Roughly one-third of the land area of the four core metropolitan area counties, for example, is protected from significant settlement or non-resource-extractive economic development

TABLE 5.1 GOVERNMENTS IN GREATER SEATTLE

TYPE

Federal	1
State (Washington)	1
Regional authorities	4
Cities	83
Counties	4
Indian reservations	5

SPECIAL DISTRICTS

Conservation	4
Cemetery	1
Cultural	1
Drainage and flood	35
Fire	89
Health-Hospital	7
Housing	9
Lake Management	9
Library	14
Park	22
Public utility	6
School	63
Transportation	4
Water and sewer	86
Total	448

SOURCE: Municipal Research & Service Center of Washington, 2009

(primarily in national parks). Several large military bases are major employers and components of the local economy, especially for Pierce and Kitsap counties.

Contemporary political relations between Indian and non-Indian communities are shaped substantially by the Point No Point Treaty of 1872 between certain tribes and the federal government. This and subsequent treaties have been subject to later revision and occasionally altered by major court decisions. Although the treaties are correctly viewed as instruments of oppression, the growth of the metropolis and subsequent federal court decisions have increased the tribes' relative political and economic clout. Reservations, especially in Kitsap and Snohomish counties, occupy attractive and valuable land and exercise a near-monopoly in the gambling industry, through more than two dozen casinos. The 1974 "Boldt Decision" recognized tribal rights to much of the salmon in "accustomed territorial waters" far beyond the reservations,[1] and granted tribes influence over development along all the major rivers of the region, as needed to protect salmon resources. Consequently, there are heightened hostilities toward Indian peoples from non-Indian fishers and others who see themselves as negatively impacted by the increased economic clout of Indians.

Washington State owns large areas of forest land operated as a productive resource for the support of schools. Much of it is protected from significant human settlement. In addition, the state exercises great influence over the economy and everyday life through financing and regulation of major highways and bridges, including ownership and operation of the ferries, which are part of the state highway system. More broadly, the state influences urban development through a maze of legislation, the most prominent of which may currently be the Shorelines Management Act, which regulates lands adjacent to waterways and lakes and protects wetlands, and the Growth Management Act, which requires and empowers local governments to control urban settlement.

As in most of the United States, counties are historically responsible for the justice system, including courts

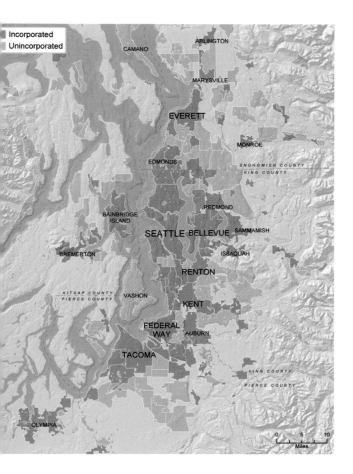

Incorporated
Unincorporated

5.1 Incorporated and unincorporated places
in the Seattle region.

and prisons, and for local roads. Larger counties, including all four in Greater Seattle, gradually gained more power over time, including land-use and building-regulation power over the unincorporated parts of the counties. Counties have some responsibility for the poor and for public health. Counties have gained even more power in recent decades, a point to which we will return after discussing special districts and authorities.

Cities were traditionally the only governments with fairly broad police powers over the citizenry and over the nature of settlement. Zoning powers had become fairly strong by 1956, when Seattle adopted its first comprehensive plan of development. Seattle and Tacoma are unusual in that they have extraterritorial ownership and power in the form of public power utilities (Seattle City Light and

Tacoma Power), with Seattle City Light's authority extending even out of the state into Idaho. The Seattle region has eighty-two incorporated places.

From the beginning, much urban settlement in the Seattle region occurred largely unregulated and outside the control of cities. By 1955 hundreds of thousands of people lived in fairly dense urban settings, but they refused to incorporate or be annexed. The state responded by creating special purpose districts for unincorporated areas, usually for fire protection, water, sewers, or combinations of these, and even for mosquito control. Special districts are taxing-authorities, and they have elected boards. Note that school districts are totally distinct systems with unique boundaries, even if they sometimes share the same name with a city, like Seattle or Bellevue.

The Growth Management Act envisions the gradual elimination of urban unincorporated territory, and incentives were added to hasten incorporation and annexation. When an area becomes incorporated, the relevant special districts may be annexed into the city or may remain independent. This trend toward incorporation, and the intentional blurring of distinctions between special districts and municipalities that it entails, challenges a traditional political culture in the region of minimalist government and local control. The result has been an increasing cultural-political divide between unincorporated rural areas and populations—particularly in eastern Snohomish, King, and Pierce counties—and incorporated areas in and around the cores of the metroplex (Seattle, Tacoma, Everett, and their immediate suburbs). In recent decades, this divide has replaced a more traditional one between the region's core cities and their suburbs. The Growth Management Act added to the effective powers of cities, counties, and regional planning bodies like the Puget Sound Regional Council.

Given the long-standing fear of big government, the preferred approach for dealing with regional externalities and interdependencies in Washington has been to create special authorities, rather than to follow any general-purpose regional governance. The most important such

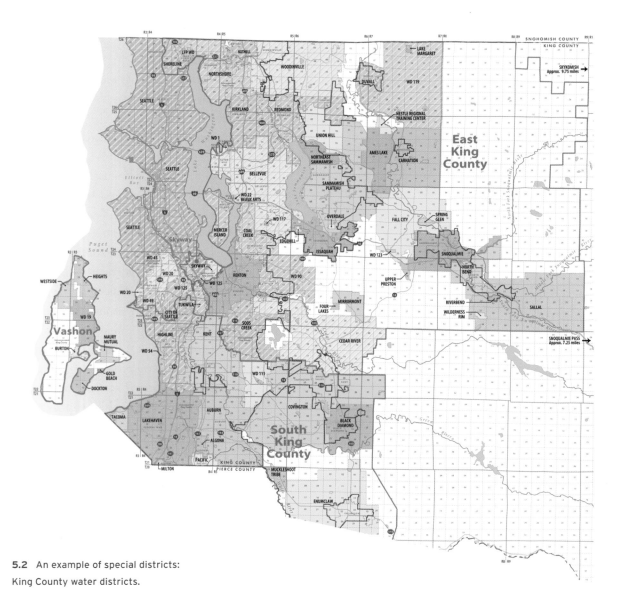

5.2 An example of special districts:
King County water districts.

entities include authorities for trade and air travel (ports), for human waste (sewerage), for solid waste (waste management), for bus and rail transportation (Sound Transit), for pollution (air and water quality entities), for water and water management, and for public power. The ports of Seattle, Tacoma, Everett, and Bremerton own and operate significant facilities and have elected boards. The Port of Seattle also runs the Seattle-Tacoma International Airport.

The history of regional governance actually starts with sewerage. By the late 1950s, the volume of mostly untreated human waste dumped from dozens of separate cities into Puget Sound and Lake Washington reached crisis propor-

tions, and the state was forced to close popular beaches. After long and complicated negotiations, the state created the Metropolitan Corporation of Seattle and forced it onto a largely unwilling populace. "Metro," as it came to be known, was run by an intergovernmental appointed board. The clean-up of Lake Washington turned out to be a popular success, and King County rewarded Metro with control over bus transportation in 1973.

Self-styled "good government groups," such as the League of Women Voters and Seattle's Municipal League, wanted regional governance for coordination of land-use planning, transportation, and overall development. Coordinated regional governance proved too threatening for citizens and existing governmental entities, and Metro was sued by a group of citizens from unincorporated King County on the grounds that its elected representatives were biased toward incorporated areas and against populations in unincorporated areas. The citizens won, but rather than having the grounds for representation changed, Metro was simply merged by the federal court into King County, making King County one of the more powerful county governments in the United States.

ELECTORAL GEOGRAPHIES

Larry Knopp and Richard Morrill

The Seattle region has a reputation as Democratic and "liberal," but this characterization is relatively recent and does not apply to the entire metropolis. Moreover, there always has been marked variation in partisan and other political allegiances and in social and political values. One indicator of these allegiances and values is election results. The bottom level of the electoral system is the political precinct, consisting of a few hundred voters and designed to nest within the whole complex of levels of government. "One person, one vote" requirements apply to most elections, so that subentity electoral districts must be revised after each decennial census. In the state of Washington, and particularly in King, Snohomish, Pierce, and Kitsap counties, such redistricting is normally carried out by a nonpartisan body or commission, which in practice has tended to maintain the status quo and protect incumbent officeholders.

The city of Seattle is unusual in electing council members at large. Voters have accepted this anomaly on the grounds of avoiding "parochialism," but they also view it as having weakened neighborhoods and reduced ideological (but, interestingly, not ethnic, racial, or cultural) diversity on the City Council.

Democrats do currently dominate the region, given results for state officers and legislative and congressional representatives. This has not always been the case, however. Throughout much of the twentieth century (from the New Deal into the 1970s), the area was quite evenly divided, with the balance of power lying in suburban King County. Seattle's "north end" and its northern and eastern suburbs were traditionally white collar neighborhoods that tended to vote Republican, while Seattle's more blue-collar "south end" and the suburbs around Boeing plants in Renton and south Seattle tended to vote Democrat. Hence, close votes for president in 1960 (Nixon), 1968 (Humphrey), and 1976 (Ford) saw King County delivering the winning margins in all three cases. The three more solidly blue-collar neighboring counties, meanwhile, with their smaller but industrial core cities of Tacoma, Everett, and Bremerton, voted consistently Democrat in all three elections.

By the late 1980s, demographic and cultural change had changed this pattern significantly. Seattle's north and south ends became less distinguishable from one another as the city as a whole gentrified (due in large measure to the rise of Microsoft and its hundreds of spin-off companies), while the city's suburbs became more ethnically and culturally diverse. Neighboring Snohomish, Pierce, and Kitsap counties, meanwhile, experienced significant middle-class spillover population growth (due particularly to skyrocketing real estate values in King County).

The result has been the establishment of King County as a strongly Democrat-voting "base," with Snohomish, Pierce, and Kitsap counties moving into "swing" categories, with social issues and land-use and growth management issues trumping class as a basis for voting (on *both* sides of the class divide). In 1988, when the state as a whole voted narrowly for Democrat Michael Dukakis in what was nationally a solid victory for George H. W. Bush, King County voted solidly Democrat, while Snohomish, Pierce, and Kitsap counties split narrowly 2 to 1 for Bush. Since then, the Seattle region has coalesced around Democrats, particularly in presidential elections.

Figure 5.3 shows the results for the presidential vote in 2008, won overwhelmingly by Barack Obama. Close results for governor in 2004 and 2008, and for attorney general

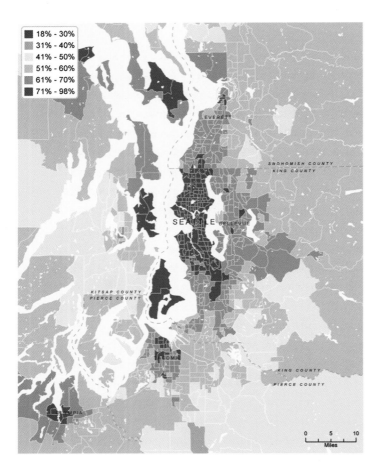

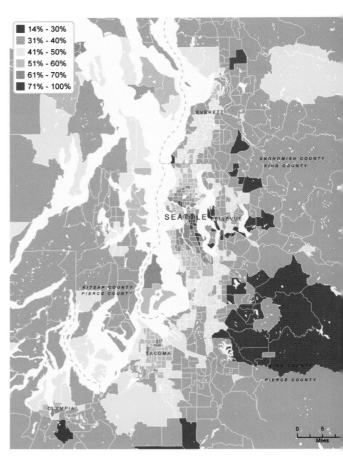

5.3 Percent for Obama in census tracts, 2008.

5.4 Percent for Ladenburg in census tracts, 2008.

in 2008, show that the region is not monolithically left-leaning, however. While the 2008 election was a triumph for Democrats, it was also surprisingly good for Republican Rob McKenna, the victorious candidate for attorney general and one of only two Republican officials elected statewide. The maps reinforce the simple and strong geography of three worlds: the city of Seattle, suburbia (including inner urban Pierce and Snohomish counties), and exurban/rural environs. Rural areas, except for anomalous Vashon and Bainbridge islands (which are in many ways more suburban in character than rural, despite population densities), tend to vote Republican (although Obama weakly carried quite a few areas); suburbia, moderately Democratic to moderately Republican (McKenna); and Seattle, well off the chart in its overall Democratic orientation (but even in the city there

was defection to McKenna). The Seattle region exemplifies here a broader national trend, with inner urban areas giving extreme margins to Obama, while rural, small town areas were much more likely to resist the "call for change."

The correlation between the vote for Obama and for the Democratic candidate for attorney general (Pierce County Executive John Ladenburg) was a rather high 0.90. But this implies that as much as 24 percent of voters chose Democrat Obama and Republican McKenna. Seattle voters (over 75 percent) went strongly for Obama, while only a few precincts in Pierce County (around the University of Puget Sound) and Snohomish County (the Tulalip Reservation) were as high. In Seattle, the highest shares tend to reflect the distribution of the highly educated as well as the distribution of African Americans. Moderately high shares for Obama (60–75 percent) dominate suburban King County, the city of Tacoma, and the denser, older suburban southwest Snohomish County. Obama weakly won much of the next geographic tier (north central King) much of outer suburban and exurban Snohomish, and much of suburban Pierce—but all these areas were fairly strongly for McKenna, and many went marginally Republican for governor. McCain carried much of exurban and rural Pierce County and southeastern King County (more populist, less educated) and scattered exurban and rural precincts in Snohomish County.

For attorney general, the pattern is similar but the results different, because of a relative Republican shift almost everywhere. High Democratic margins (over 60 percent) are confined to central and southern Seattle, Vashon Island, and several areas of the city of Tacoma, reflecting John Ladenburg's popularity as Pierce County's executive. Ladenburg more weakly carried north Seattle as well as inner southern suburbs, less familial areas along a major state highway corridor in Snohomish County, and south suburban Tacoma. But McKenna dominated almost all of Seattle's eastern suburbs, eastern and far southern King County, most of Snohomish County, and most of outer suburban, exurban, and rural Pierce County.

Overall these patterns suggest that Republicans cannot currently carry Seattle and that a charismatic Democrat (e.g., Obama) can win some small town and rural areas, but with little or no "coattail" capacity. On the other hand, suburbia, with the large majority of the region's population, remains a partisan battleground and offers an opportunity for moderate Republicans (if they can win their party's nomination).

The region exhibits a broad correlation between Democratic partisanship and more "modernist" (or liberal) voting on social issues. But there are differences, exemplified by two referenda on social issues. The November 2008 Initiative 1000 (popularly known as the "Death with Dignity Act") essentially legalized physician-assisted suicide under certain conditions. Referendum 71 (November 2009), meanwhile, affirmed a legislatively adopted domestic partnership law (informally known as "everything but marriage"), which was largely perceived as a gay-rights measure and was brought to the voters by a coalition of conservative forces seeking repeal of the legislature's action (strangely, the process for repealing a legislatively enacted law in Washington State entails petitioning for a referendum on whether or not to affirm it). Finally Initiative 1033 (November 2009) would have severely constrained long-term revenues for state and local governments. The almost 58 percent statewide victory for Initiative 1000 was surprisingly strong considering the seemingly forceful opposition and the fact that the measure had lost in the 1990s.

In both Snohomish and Pierce counties support for I-1000 in the urban cores (and older suburbs) was mainly positive, but consistently lower than support for Obama. But in the outer suburbs, exurbia, and especially the far rural areas, support for I-1000 was often quite strong, even in many areas that supported McCain. This pattern may reflect the older populations (who tended to vote yes) in more distant rural areas, as well as in the inner cities. Or it may reflect a libertarian impulse that is strong in rural areas. The overall weaker support for I-1000 in Pierce than in Snohomish (or King) County probably reflects a less affluent, less educated, and more military-dependent population. Seattle's support for I-1000, while quite high, was markedly less than for Obama, while far suburban, exurban, and rural area support was much stronger. The lower levels of support in south King County than in the eastern suburbs is similar to what was noted for Pierce and Snohomish counties.

In the city of Seattle, while areas with high concentrations of the elderly were highly supportive of I-1000, the highest levels of support were ironically in the area of highly educated, often younger nonfamily populations in Capitol Hill, Wallingford, and Fremont. The only area with a majority against the initiative was the poorer area of southeast Seattle, with higher levels of recent immigrants and ethnic and racial minorities (also the case in Bellevue). In sum, the vote for I-1000 differed from that for partisan

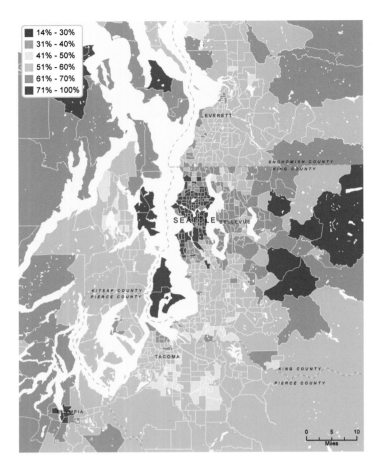

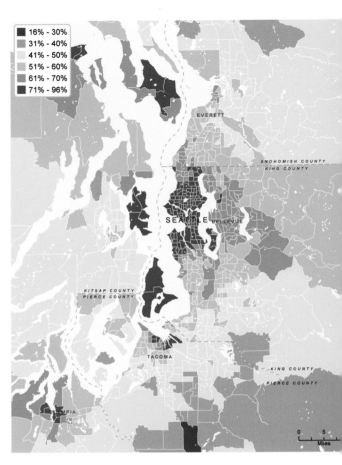

5.5 Percent for Initiative 1000 in census tracts, 2008.

5.6 Percent for Referendum 71 in census tracts, 2008.

races by the greater support from the elderly and weaker support from minority populations.

A second illustration of nonpartisan, social-issue voting is the November 2009 vote on Referendum 71, brought to the voters by opponents of a legislatively passed domestic partnership measure ("everything but marriage," but not quite). Results are shown in figure 5.6 for the Greater Seattle area. The geographic pattern is a remarkable mapping from socially liberal to conservative—almost a cliché. With few exceptions, the strong yes vote almost coincides with the city of Seattle, with a drop off only in the far south of the city. Moderate support, 50 to 70 percent, characterizes most of King County suburbs and exurbs, especially in the more professional east, while a majority against Ref. 71 prevailed in some obvi

ously urban as well as exurban and rural southern and southeastern King County precincts, dominated by craft and laboring occupations, manufacturing, construction, transportation, and utilities.

Analysis of data at the census-tract level shows a high correlation of support for Ref 71, with high shares of transit users, people in nonfamily households—especially same sex households and younger adults aged 20–39—and those in areas with high numbers of professional and financial, insurance, and real estate occupations. Correlations were highly negative with shares of husband-wife families, the population under 20 (a surrogate for the presence of children), persons in manufacturing, craft, and laboring occupations, those with only a high school education, and those who are single-occupancy vehicle drivers. These are similar but not identical to the results for the presidential election (the correlation of Ref 71 and the vote for Obama was 0.91).

Elections, of course, are only one manifestation of political power and decision making in the city. In the next section, we explore a rather different kind of political geography.

THE PARADOXES OF SOCIAL CONTROL IN SEATTLE

Steve Herbert

Just north of the popular Pike Place Market is Victor Steinbrueck Park. The park is less than an acre and consists mostly of a grassy lawn and several benches. Its signal feature is the view it provides of Puget Sound and the Olympic Peninsula. On a sunny afternoon, it is a wonderful place to relax and appreciate the natural beauty that surrounds Seattle. Yet many people avoid this park. It is frequented by the homeless, many of whom like to share alcohol and recreational drugs. Its long-standing reputation as the site of low-level criminal activity leads many to find another place to enjoy their Market-purchased lunch.

In response to the dynamics in places like Victor Steinbrueck Park, the City of Seattle has crafted a range of social control mechanisms. These mechanisms are designed to compel individuals to abide by general norms and hence to behave in an orderly and predictable fashion. One of them is a 1997 ordinance that enabled "park exclusion orders," which empower police officers to exclude any individual from a park who violates any law or rule. This means that one can not only be issued a citation for, say, being in a park after hours but also be prohibited from reentering the park. The length of the prohibition varies, from a week to a month to a year, depending on one's past history of exclusions. Victor Steinbrueck Park witnesses the largest number of such exclusions of any park in Seattle.

About a mile due east sits a building that exemplifies a different response to those who people the park. The Downtown Emergency Services Center, a nonprofit organization that provides low-income housing and an emergency shelter, operates 1811 Eastlake, a shelter dedicated to chronic alcoholics. Residents are allowed to drink on the premises of this "wet" housing, which sets 1811 apart from most supported housing facilities that typically demand sobriety.

The logic of wet housing is simple. Chronic alcoholics, by definition, are likely to continue to drink; therefore, better for them to do so indoors in an environment that is safe, warm, and permanently staffed. In such an environment, residents are less likely to harm themselves and others. Further, the facility provides continual opportunities for residents to address their alcoholism through various forms of treatment. The logic appears to work: a recent evaluation of 1811 Eastlake showed that residents drink less than before entry. They also land in jail and the emergency room far less frequently, at a considerable cost savings to taxpayers.[2]

The progressive philosophy behind 1811 Eastlake is one that is commonly associated with Seattle. Yet Seattle is simultaneously a leader in techniques of spatial exclusion. Its reputation in this regard was forged most notably in the 1990s, when City Attorney Mark Sidran spearheaded efforts to implement a wide range of so-called "civility codes." These were ordinances aimed at common behaviors engaged in by the homeless and others who spend a considerable amount of time in public spaces. The new laws made it illegal to sit or lie down on a downtown sidewalk, to panhandle aggressively, and to camp in parks. The parks exclusion ordinance was part of the Sidran-led effort. The civility codes are now combined with other mechanisms to increase the police's ability to make arrests. The most notable of these are criminal trespass and off-limits orders. Criminal trespass arrests rely upon a preexisting formal agreement between the Seattle Police Department and a property owner where the police may act as a trespass agent without the specific permission of the owner in any given case. So if an officer sees a person "without legitimate purpose" on a property where a contract is in operation, that officer can "trespass admonish" the person. That "trespasser" cannot reappear on that property for the next year.

Violation of the trespass admonishment leaves a person vulnerable to arrest for criminal trespass.

Off-limits orders operate on a similar logic, and are applied consistently to those who are charged with a drug- or prostitution-related offense. Such individuals will typically be issued with either a "Stay Out of Drug Area" (SODA) or "Stay Out of Areas of Prostitution" (SOAP) order. SODAs and SOAPs are predesignated areas with an established history of street-level drugs or prostitution. Any individual caught by the police in a prohibited area is subject to arrest. Because SOAP and SODA orders are increasingly common, arrests for violations show a steady increase in the recent period.[3] The size of the areas covered under these laws has also increased, especially for SODA zones. As figure 5.7 indicates, about half of the city's area is currently included in one of these zones.

Seattle thus appears a paradoxical city when it comes to social control. It hosts new and vigorous forms of exclusion *and* a nationally recognized inclusionary approach to a particularly disadvantaged population. Any effort to explain this paradox must start with the recognition that Seattle, like all places, is strongly affected by larger social, economic, and political forces. The U.S. urban homeless population increased dramatically in the 1980s, largely because of the Reagan administration's wholesale withdrawal from federal public housing programs. At the same time, many U.S. cities saw an increase in gentrification, whereby undervalued downtown properties were converted into housing for upscale professionals (see chapter 6). Seattle, for example, saw a decrease of about half of its single-room occupancy (SRO) housing between 1970 and 1985.[4] This represented a loss of more than 15,000 units of low-income housing.

The influx of more affluent downtown residents generated ripple effects. While gentrification and urban redevelopment brought more people downtown, its business community grew concerned that the increased presence of the homeless would deter sidewalk traffic. As a consequence, they pressed for the civility codes and other "projects of reassurance," to use Timothy Gibson's term.[5] Such measures were intended to send a signal to the broader

SODA 2005

5.7 Stay Out of Drug Areas (SODAs) in Seattle.

community that downtown could be a safe and pleasant place to shop and play.

These projects of reassurance were legitimized by the "broken windows" theory. This term was coined by two criminologists, James Wilson and George Kelling,[6] who argued that homeless people, street walkers, and public inebriates constituted a form of "disorder" that eroded an area's quality of life and increased the prevalence of more serious crime. For this reason, they argued, the police should focus attention on these individuals. Seattle's civility codes were intended to give the police the means to do so, by granting them increased authority to make arrests of those described as disorderly. Although the broken windows logic—that disorder today leads to more serious crime tomorrow—does not stand up to empirical scrutiny,[7] it became quite popular and remains so to this day. Seattle police, for example, commonly cite it in public forums as justification for their social control practices.[8]

All of these factors—a rise in the visible homeless, increased gentrification and downtown redevelopment, and the ubiquity of the broken windows discourse—help to explain why Seattle has intensified the social control net in recent years. Yet this intensification does not appear to generate many notable positive impacts. Take the various forms of exclusion now practiced in Seattle. Interviews with homeless people who are subjected to these exclusion orders make plain that their already difficult lives are complicated further when they are banned from particular places. They find it harder to maintain connections to friends, family, service providers, and possible work opportunities. The pull of these connections means that individuals routinely violate their exclusion orders; many feel they have no choice. Yet even if their behavior remains unchanged, they carry the stigma of marginalization.[9] If they are arrested, they usually get but a few nights in jail.

What exclusion *does* provide is a means by which police, prosecutors, and politicians can continue with "projects of reassurance." They can try to convince members of the business community and other concerned citizens that something is being done to address their concerns about "disorder." This is an expensive proposition, however, because the cost of jail time is significant.[10]

Here again 1811 Eastlake provides a compelling contrast. Although it costs about $13,400 per client to run the shelter, those same clients were costing taxpayers about $43,000 prior to their admission. To reduce jail nights and emergency room visits is to reduce public expenses by a significant amount. This cost-benefit analysis, according to the director of DESC, is what explains the support his organization receives from the downtown business community (despite its general enthusiasm for broken-windows policies).

One of the more pernicious consequences of the discourse of broken windows is the association it reinforces between disorder and the homeless and other publicly visible disadvantaged populations. Symbolized as disorderly, the homeless become less easily seen as rights-bearing individuals who deserve compassion more than castigation. Practices that formally exclude them from public space are thereby much more easily justified.

By comparison, 1811 Eastlake evidences a different public impulse, and it needs to be kept in mind by all of those who experience fear in Victor Steinbrueck Park. It represents a geography of hope and care, a strong statement about collective responsibility. Its residents are more than an example of "disorder"; they are struggling individuals who need assistance. It represents the other side of Seattle's urban politics, and stands as a symbol of the better sort of city that Seattle can be.

WHOSE PUBLIC SPACE? THE POLITICAL GEOGRAPHY OF SKATEBOARDING

John Carr

While it may seem odd to discuss youth recreation facilities in a chapter devoted to serious political analyses, political geography has long been concerned with the ways that political struggles inform the creation of public spaces. In Seattle, debates over public skateparks have exposed much broader political and cultural tensions about where young people should be in the city, and who should speak for them in the political system. In turn, these tensions have revealed deep-seated ambivalence about age and gender.

Faced with the disappearance of popular, although often illegal, street terrain through the 1990s, Seattle's skaters found themselves at the forefront of an emerging national effort to claim permanent public spaces for skateparks. While a group of young skaters tried creating a North Seattle skate area in the 1990s, the movement gained momentum only once a number of older skaters in their twenties and thirties, including veterans of a pioneering guerilla project under Portland's Burnside Bridge, became involved. In response to their efforts, the city gave these advocates permission to build a temporary skateboard park in the parking lot of an abandoned Safeway store that had been purchased for a future Parks Department project.

The involvement of older skaters, however, shifted both the focus and the constituency of this nascent movement away from young people. These older advocates came of age during the 1980s, a time when "transition" skating had been much more popular than street skating, which has been favored by younger skaters since the early 1990s.[11] Reflecting this shift, the skatepark organizers insisted that the Ballard park include a large, concrete transition feature, in addition to the temporary wooden street-style ramps originally planned. The first Ballard Skate Park was completed in 2002, utilizing only volunteer labor and a budget composed of private contributions and matching city funds. As built, the centerpiece of the park was the "Ballard Bowl" (fig. 5.8), an approximately 1,800 square-foot, pool-like structure. Working without city supervision, the volunteers created a facility that developed a national reputation for excellence.

At the end of 2003, the city announced that it would demolish the Ballard Bowl to make way for the Ballard Commons, a passive-use park. Mobilizing around the threat to the bowl feature, a number of organizations, including the Puget Sound Skatepark Association (PSSA) and Parents for Skateparks sprang up overnight, largely led by professionals—including a number of Microsoft and other dot-com employees—in their late twenties, thirties, and early forties, but also including younger street skaters. Organizing under the slogan "Seattle Hates Kids," activists waged a successful media campaign that included such newsworthy events as a skate jam/fundraiser, and a "thousand skater march" from the Seattle Space Needle to City Hall.

Previously unexposed tensions between the majority of young street skaters and older activists emerged, however, once Seattle Mayor Greg Nickels decided to back the presence of a permanent skate facility in Ballard. After the city rejected the original location of the skatepark, older advocates managed to achieve a compromise by which the street-skating features would be eliminated, but the transition-oriented Bowl feature would be rebuilt in a different section of the park. This agreement created a schism within the movement that ultimately transformed its aims, as well as the nature and locations of public skateparks throughout Seattle.

The original Ballard Bowl was by most standards an unappealing site. Located in an abandoned semi-industrial site between an alley, a supermarket trash compactor, and a disused loading dock, the Ballard Bowl felt gray, dirty, and inhospitable to outsiders. In the warm summer months, the scent of rancid fried chicken in the adjoining supermarket dumpster pervaded the entire area. Even so, many older skaters valued this setting as more authentically urban, transgressive, and "real." By creating a barrier to those uncomfortable in such an unwelcoming environ-

5.8 Original Ballard Bowl.

ment, the Ballard Bowl served as a de facto "clubhouse" for a self-selecting group of predominantly older and male skaters, many of whom spearheaded the effort to save a place for skateboarders in Ballard.

In contrast, a small but vocal number of activists (largely parents) who had become involved in the Ballard movement vehemently argued that skateparks should actually serve young people and their families. For some of these advocates, the Ballard compromise amounted to adults "selling out" young people. They envisioned skateparks as places for city kids who did not want to play organized sports. And this entailed finding spaces that were as clean, safe, orderly, and accessible to male and female children as a soccer field or softball diamond. For example, when Lower Woodland Park (fig. 5.9) was announced as the location for a new skatepark, parent-advocates convinced the recently created Skatepark Advisory Committee to the Parks Department (known as SPAC) to object because the originally identified site within the park was unwelcoming to families and invited unsupervised, illicit mischief and antisocial behavior by teen skaters. In response to these concerns, Parks shifted the location of the planned facility approximately 100 yards closer to the street and neighboring residential district, until NIMBY ("not in my back yard") objections pushed the facility to a compromise location.

The shift to a playground model—and away from more adult-oriented clubhouse approaches—had profound implications for both skatepark politics and locations. As leaders of the PSSA and SPAC focused increasingly on conventional city politics—with its emphasis on diplomacy and compromise—many of the older Ballard activists who preferred using oppositional, media-intensive tactics became disenchanted. Within a few months of Parks' decision to "move" the bowl feature, for example, a number of Ballard protest organizers disavowed constructive engagement with a Parks Department they deemed untrustworthy and incompetent. And many older transition-terrain proponents dropped out once it became clear that their interests were not foremost on the agenda.

By rejecting the pursuit of clubhouse-type sites,

5.9 Lower Woodland Skatepark, street terrain area.

youth-oriented skatepark advocates also changed both the nature of the political struggles over those sites and the resulting sites themselves. Because the playground model calls for existing recreational space to be repurposed for skateboarding, getting skateparks built means overcoming passionate resistance from existing park users. Almost every proposed skatepark site in Seattle has been met with opposition from nearby residents concerned about the loss of green passive-use space, over-programming of parks, and—above all—worries about noise, litter, and crime being introduced to their neighborhoods by (mostly) male teenage skateboarders. As one typical public comment in opposition to a proposed site in Cowan Park argued, "Why anyone would want to bring noisy skateboarders here . . . is a mystery. It's very sad for someone like myself after the countless hours I worked to dump the druggies for the use of civilized people only to be faced with noise, candy wrappers, coke cans, needles, and concrete." In turn, the city has tended to capitulate to the demands of the more powerful and wealthy neighborhoods that have objected to potential skatepark locations. Thus, when Seattle created its 2007 city-wide master plan for future skateparks (fig. 5.10), proposed sites in neighborhoods like Cowan Park and Myrtle Reservoir were excluded, while the remaining

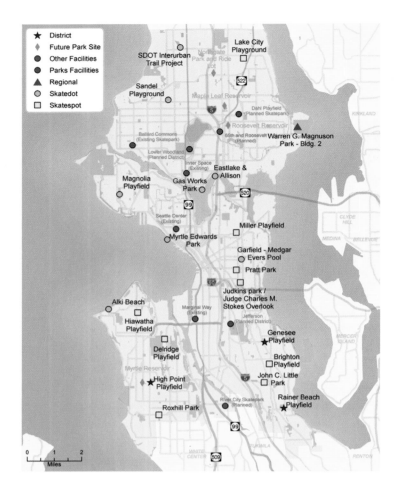

5.10 Sites identified by the 2007 skatepark master plan of Seattle Parks and Recreation.

twenty-six locations were either in neighborhoods that are at or below Seattle's median household income level or far from residential areas.

Notwithstanding the resistance of many neighborhoods, the cost to the advocacy movement, and the heightened political stakes, the youth-oriented playground model has gained acceptance from city government and appears to be leading to the promulgation of a number of new skateparks. A variety of projects have recently opened, are under construction, or are in the planning phase on the Seattle Center campus, in West Seattle's Delridge Park, in North Seattle's Dahl Playfield, and at the Crown Hill School site in Central Seattle.

So what political geography lessons can we learn from this study? First, the political debates over public skate-parks in Seattle brought to light society's powerful but often latent ambivalence about the presence of young men and boys in public spaces. Adults involved in these politics typically agreed that young people needed skateparks; few wanted those facilities in their back yard.

Second, this ambivalent perspective on young people echoed through city-neighborhood power relations. Seattle allowed the matter to be settled according to the dictates of that power, allowing every side paradoxically to "win." Powerful neighborhoods that were effective in organizing against the threat of a nearby skatepark won by having politically "hot" sites either moved or removed. In contrast, less powerful and organized residents won by having these facilities located in or near their neighborhoods (regardless of whether they wanted them).

Third, the tendency of adults to speak on behalf of young people in the political system came with important costs and risks. Even the best-intentioned adults may not appreciate the difference between their own goals and needs and those of the young people for whom they purport to speak.

Thus, the final master-plan of skatepark sites maps both the result and the representation of the political system's efforts to answer, unevenly and incompletely, questions as to who our young people are, whether and where they should be found in the city's public and nonpublic spaces, and what parts of the city should (and can be required to) receive the mixed blessings of youth-oriented recreational facilities.

GIS AND REGIONAL TRANSPORTATION DECISION MAKING

Timothy Nyerges, Kevin Ramsey, and Matthew Wilson

Everyone knows there's a transportation problem in the Seattle region.[12] Everyone agrees that the transportation system is in dire need of investment to address the problem. However, there is far less agreement regarding what kinds of investments to make and how to raise the funds to pay for those investments. So decision making around transportation investment has become a classic problem of political geography here. Essentially the question is: where should limited funds be spent? Even more challenging is how to engage the public in decision making when the "Seattle way" is typified by endless process, discussion, and debate.

Commonly, transportation agencies fulfill the federal mandate for public involvement through a "transportation improvement program" (TIP). This document includes a list of transportation projects and sources of funding to pay for those projects. Once a draft TIP is completed, the agencies then convene public meetings to gather public comments. In other words, the agencies first determine which projects to build and *then* ask the public if the list is acceptable. Involving the public at such a late point in the process severely limits their ability to shape the goals of the TIP, the projects to include, and the funding sources to pay for them.

The Participatory Geographic Information Systems for Transportation (PGIST) research project carried out by University of Washington geographers from 2003 to 2007 set out to develop and test an alternate way to engage members of the public *throughout* the transportation improvement programming process. It built and evaluated a Web site that combined innovative Web mapping with online discussion technologies capable of supporting the collaboration of hundreds, even thousands, of citizens in the creation of a TIP. The research integrated geographic information systems (GIS) technology, transportation, and political geography.

An Experiment in Participatory Democracy

The project team designed an online field experiment in participatory democracy. To support this experiment we created a Web site called "Let's Improve Transportation" (http://www.letsimprovetransportation.org; see fig. 5.11).[13] More than two hundred residents of the Seattle metropolitan area were recruited to participate. Each participant was asked to spend at least eight to ten hours over the course of twenty-eight days on a series of five tasks, at a

time of their own convenience using their home computer or public Internet connection. The five tasks were the sort that transportation planners and policy makers commonly perform when they are developing a TIP. However, we modified them to make them understandable to the general public and to take advantage of a wide diversity of local knowledge that participants brought to the table. Our goal was to provide enough structure to guide participants through the creation of a TIP but allow enough flexibility to encourage new ideas that do not conform strictly to planners' expectations.

To make the experiment as realistic as possible, we modeled it upon an actual transportation investment-decision problem facing the Seattle region (see fig. 5.12). Each participant was asked to imagine that he or she is a member of a large citizen advisory committee, charged with providing policy makers their recommendations regarding a (fictitious) regional transportation ballot measure. The list of proposed transportation projects and funding sources was based on the same list policy makers drew from when deciding how to compose the real-world ballot package.

The experiment was open to anybody who wished to participate. The research team made a special effort to

5.11 The LIT Web site homepage, https://www.letsimprovetransportation.org.

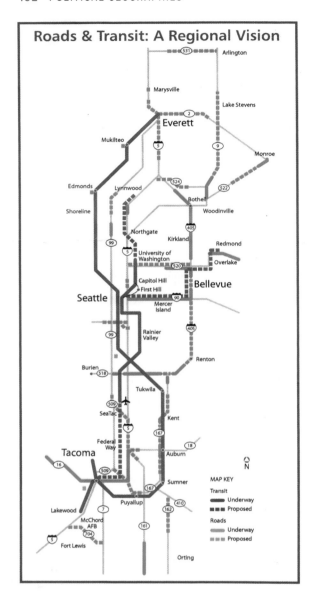

5.12 The proposed "Roads and Transit" package put before Seattle area voters in 2007. The LIT field experiment was designed as an alternative way to generate a transportation package through greater public involvement.

ensure the participant pool included representatives from all geographic areas in the region as well as harder-to-reach populations, such as lower income residents and ethnic/racial minorities. A limited number of stipends for

paid participation were offered in each of five different geographic regions, made available on a first-come, first-served basis.

THE PARTICIPANT'S EXPERIENCE

The LIT Web site included a collection of tools and information supporting analytic and deliberative activities organized into a five-step sequential process (see table 5.2). During the field experiment, the Web site's automated agenda manager allowed participants to proceed step by step during specific calendar dates as guided by a moderator.

Using the LIT Web site, participants worked asynchronously as a group at their own convenience but roughly within the same general time frame through the five-step workflow agenda (fig. 5.13). Each step included tasks for participants to complete individually, and there was a discussion area for deliberating with other participants. In step 1, participants shared information about their daily travel habits and their opinions about improving transportation. The moderators synthesized participants' concerns and grouped them into a set of themes. Participants reviewed the themes and voted on whether they agreed that the themes adequately represented their concerns. In step 2, participants reviewed and weighed different factors (or criteria) that could be considered when deciding which projects to prioritize. In step 3, participants reviewed proposed transportation projects and evaluated how closely they aligned with their preferred planning factors. Then each participant created their own transportation improvement package, selecting from a list of projects and funding sources (fig. 5.14). The Web site tracked the balance of costs and revenues to ensure that each package had sufficient funding. Next, a transportation planning specialist used a semiautomated clustering process to identify six representatively diverse packages. In step 4, participants reviewed, deliberated, and then voted on which package they preferred. Finally, in step 5, participants reviewed and endorsed a final report to agency decision makers and technical specialists.

TABLE 5.2 STEPS IN THE 28-DAY FIELD EXPERIMENT

"LET'S IMPROVE TRANSPORTATION"

STEP 1. Discuss concerns

1a: Map your daily travel

1b: Brainstorm concerns

1c: Review summaries

STEP 2. Assess transportation improvement factors

2a: Review factors

2b: Weigh factors

STEP 3. Create transportation packages

3a: Discuss projects

3b: Discuss funding options

3c: Create your own package

STEP 4. Select a package for recommendation

4a: Discuss candidate packages

4b: Vote on package recommendation

STEP 5. Prepare group report

5a: Review draft report

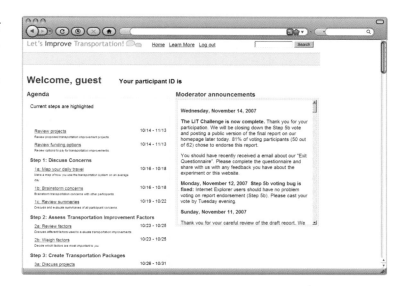

5.13 A portion of the LIT workflow agenda page with moderator announcements shown on the right-hand side of the page.

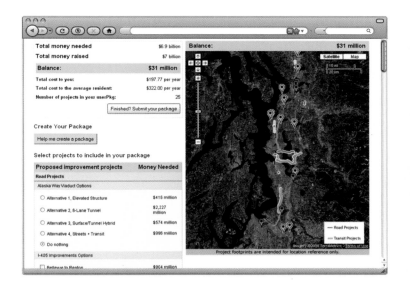

5.14 LIT participants design their own package of transportation projects and funding sources. The selected projects are displayed on an interactive map.

RESULTS OF THE EXPERIMENT

A total of 246 participants registered for the experiment. Of those registered, 153 were active at some point during the twenty-eight-day experiment and 135 contributed at least one concern, discussion comment, vote, or package (in other words, they did more than just "lurk").

Most of our participants were actively interested in the topic of improving transportation and felt comfortable using online tools. About 90 percent of those reporting said they used the Internet several times a day and considered themselves to be either experts or nearly experts in terms of their familiarity with computers and the Internet, while nearly two-thirds (65 percent) said that they had participated in online discussions before. Almost all (94 percent) of those reporting stated that they tried to stay informed about transportation issues in the Puget Sound region, and one in five participants indicated they were either affiliated with or an actual member of an organization that was active in transportation planning or advocated for transportation issues.

As might be expected for a twenty-eight-day process, we had some participant attrition during the experiment. For example, 57 participants voted on package recommendation (step 4b) and 47 voted on report endorsement (step 5b). Among those who did participate in these votes, there was a relatively high degree of consensus: 61 percent endorsed the winning candidate package (1 of 5 packages); 81 percent endorsed the final report that described the result of the decision process.

The package that 61 percent of LIT participants endorsed was notably different from the real-world, ballot-measure package in several significant ways. The first difference was funding. While the ballot measure was funded almost exclusively through an increase in the regional sales tax, the LIT package relied instead on region-wide highway tolls and other "user fees," such as a gas tax, vehicle license fee, and commercial parking tax. Second, the LIT package was much smaller, costing only $11.8 billion compared to the ballot measure which sought $17.8 billion in funding for road and transit improvements. Third, the LIT package included a number of recommendations regarding improvements to transit systems. The ballot package primarily focused on system expansion. A considerable number of people wanted more transit options than represented in the real ballot measure.

To investigate the use of GIS maps during the LIT online field experiment, we identified two key event types from the system-event log indicating when participants were browsing maps of roads or transit projects. Decoding the two event types allowed us to determine how participants were using GIS to make their decisions. Based on the event logs, there were clear differences in what participants were interested in looking at. For instance, maps of the four proposed alternative replacement projects for the Alaskan Way Viaduct were viewed much more frequently than average. One of these project maps was browsed thirty-five times over the course of the experiment, more than any other single project map. There did not appear to be any significant relationship between the number of times a project map was browsed and the amount of money needed to pay for it. Even though some proposed projects were estimated to cost upwards of $2.5 billion, such as the alternatives for replacing the SR-520 bridge across Lake Washington, these particular projects were browsed less by participants than other projects such as "Snohomish County SR-9 improvements," which were estimated to cost only $300 million.

Seventy-five participants filled out our post-experiment questionnaire. The responses indicated a relatively high level of satisfaction with LIT as a model of public participation. For example, 73 percent agreed (or strongly agreed) with the statement "I believe the LIT Challenge is an example of a meaningful and productive way in which members of the public could participate in decisions regarding how to improve our transportation system"; 88 percent agreed with the statement "I believe my own understanding of the transportation system, and ways to improve it, can be enhanced through discussions with other members of the public who may have different perspectives than my own." Slightly less (75 percent) agreed with "I am interested in having these kinds of discussions." Finally, we were curious to discover that their agreement continued to fall—to 66 percent—when asked if they were "interested in having these kinds of discussions on the Internet."

Preliminary analysis of semi-structured interviews mirrored the generally positive responses seen in the post-experiment questionnaire. Overall, interviews with twenty participants reflected that, even given certain technical glitches, online discussions were both useful and convenient—as an educational tool, as outreach, and as a means of venting frustration. Several even commented that they learned more through discussions with other participants than they did through reading the "expert" analysis of transportation projects available on the LIT Web site. Participants who reported these optimistic impressions about the deliberations also tended to be some of the most active in online discussions. Therefore, in future research we plan to examine in more detail the differences in experience and satisfaction between those who are most active in discussions and those who did not actively contribute to discussions. Although we engaged a large, diverse group, we cannot claim that it was representative of the entire population in the region.

REFLECTIONS ON OUTCOMES

Despite the anticipated drop-off in participation, we conclude that the moderator's role worked well in organizing concerns, relating them to the issues at hand, and synthesizing them. However, our use of keyword tags—a feature common on social networking Web sites such as Flickr—was not entirely straightforward for some participants. Allowing participants to see how moderators categorize their keyword tags and concerns and to contest this, if they disagree, might be an alternative approach.

The five packages identified by the computer-based clustering process each had idiosyncrasies unique to the preferences of individual participants. Given that each of these packages was designed by a single participant, it was perhaps unsurprising that each had a different geographic bias in project locations. This led to some frustration by participants about the quality of the five packages available to endorse in step 4. This outcome appears to be an inevitable by-product of the design of our candidate-package selection methodology. A better solution, we believe, is to generate a new "representative" package based on the overall characteristics of the individual packages within each cluster. Alternatively, an overall average package could have been developed to promote fairer geographic distribution of projects or consistency in the application of tolls.

In the end, we considered the experiment a success because a package was recommended. However, the 2007 ballot measure was soundly rejected by voters at the polls, a stark contrast to the recommendation in the experiment. That ballot measure was based on several years of negotiations by political leaders throughout the region regarding the composition of the package, with heavy emphasis on roads. Involving interested voters early in the package-creation process appeared to be a significant aid in gaining voter approval for transportation funding measures. People would likely prefer to pay for what they want, rather than for what they do not want. Let's Improve Transportation represents one innovative strategy for enabling the assemblage of such packages by diverse stakeholders in the regional community.

THE ALASKAN WAY VIADUCT, CLIMATE CHANGE, AND THE POLITICS OF MOBILITY

Kevin Ramsey

Between 2001 and 2009, the most contentious transportation infrastructure investment decision in the Seattle region was the Alaskan Way Viaduct, the aging elevated highway that runs along the downtown Seattle waterfront (fig. 5.15). For over eight years after the already crumbling structure was damaged in 2001 by the Nisqually earthquake, state and local elected officials debated how to replace it. Seattle Mayor Greg Nickels and many downtown property owners, businesses, and developers supported replacing the Viaduct with a cut-and-cover tunnel. Meanwhile many state elected officials, industries south of downtown, and outer-neighborhood residents supported building a new, larger elevated highway. Some business owners and activists went so far as to accuse tunnel supporters of "class warfare" for their willingness to put a key freight corridor and working-class jobs at risk in favor of improving views, especially for the downtown elite. A third, smaller group called for simply removing the Viaduct entirely and investing in the existing surface-street grid, public transit, and environmental amenities.

The Viaduct debate revealed a great deal about Seattle's "politics of mobility": political struggles over the configuration of urban space and the kinds of mobility that this configuration should facilitate. Spaces optimized for one form of mobility—for instance, "automobility" privileges personal automobile travel—can undermine other kinds of mobility, such as walking, biking, or rapid transit. Urban transportation planning is not simply a matter of determining the greatest efficiency. It also entails normative judgments (often tacit) regarding which kinds of urban

5.15　Alaskan Way Viaduct.

lifestyles and urban economies should be encouraged or discouraged by infrastructure.

During the debate, concerns about climate change helped shape the politics of mobility in Seattle. In 2005 Mayor Nickels gained international recognition for spearheading the U.S. Conference of Mayors Climate Protection Agreement. Mayors who signed the agreement committed their cities to achieving Kyoto Protocol greenhouse gas reduction targets—7 percent below 1990 levels—by 2012. Advocates for a "surface/transit" Viaduct replacement option in Seattle quickly worked to highlight the contradiction between building a new highway on the waterfront and Seattle's new goal to significantly reduce greenhouse gas emissions from transportation. Although it was initially rejected by both transportation planners and the mainstream media, the surface/transit campaign gained significant momentum in 2006, when state and local elected officials failed to reach agreement on a Viaduct replacement option.

Climate change grew as a prominent theme in statements of support for the activists' surface/transit option in the local media. Major newspaper editorials and public statements by elected officials began calling for transportation agencies to predict the greenhouse gas emissions of differ-

ent Viaduct replacement options. More broadly, they called for a fundamental shift in how transportation planning is approached in Seattle. Specifically, they questioned the dominance of automobility and asked that Seattleites re-imagine how people and goods can be moved through the city.

This discursive shift against automobility and toward climate-change amounted to a challenge to the dominant approach to transportation planning practiced by the Washington State Department of Transportation. Key assumptions regarding the need continually to replace or expand vehicle capacity, on roadways, for example, were no longer seen as self-evident and outside the boundaries of public debate.

While the activists succeeded in expanding the politics around the Viaduct question, the debate was still far from resolved. In 2007 a high-profile, stakeholder advisory committee was convened by state and local elected officials to evaluate new Viaduct replacement options. Surface/transit advocates serving on the committee asked agency leaders to translate their demands into quantitative measures for evaluating Viaduct replacement options. In doing so, however, they inadvertently ceded the terms of debate to a discourse on automobility. Agency planners simply adapted existing transportation-modeling techniques to predict the future "carbon footprint" of the regional transportation system after proposed Viaduct replacements are built. These predictions use essentially the same conservative assumptions about the ever-increasing demand for automobile travel that have informed the work of transportation planners for decades. Consequently, they cannot envision a future where automobility is not the dominant discourse in transportation planning. They cannot imagine how life in the region might be organized differently than it has in the past. The automobility assumptions built into the model were reflected in the agencies' findings: All proposed Viaduct replacement scenarios—including five highway and three surface/transit options—were predicted to increase greenhouse gas emissions in the Seattle region to 14–15 percent above current levels by the year 2015. Furthermore, climate change per se was no longer a reason to fundamentally rethink Seattle's transportation system or the institutions that maintain it.

In the end, state and local elected officials opted to build the most expensive option: a bored tunnel. Decision makers hoped that this option would address concerns about construction impacts by digging a new highway deep beneath downtown while allowing traffic to continue flowing on the old Viaduct. Surface/transit advocates continue to oppose the tunnel; however, they no longer invoke climate change in their arguments. One way to interpret this outcome is to conclude that the political potential of the "climate crisis" as a call for urban and institutional transformation was greatly diminished. At any rate, once the future carbon footprint of different Viaduct replacement scenarios essentially became fixed and known through an open and relatively conflict-free process, the issue could no longer serve as an inspiration to create a different kind of city.

So, one important lesson from this political geography is that quantitative modeling is only as good as the assumptions built into it. Yet this logic of "predict and provide" persists within transportation agencies, systematically reproducing the dominant paradigm of automobility, in Seattle and elsewhere. Placing limits on local greenhouse gas emission can appear to environmentalists as a silver bullet in their quest to undermine the persistent dominance of automobility in structuring urban policy decisions. However, there are significant limitations to this strategy. Instead, activists, planners, and scholars must directly and persistently challenge the system of automobility, not merely call for the measurement of its effects. In other words, we must directly address the many social, cultural, economic, and environmental factors that both perpetuate automobile dependency and normalize its desirability. A good place to start would be in neighborhoods most dependent on the Viaduct for automobile travel.

CODING COMMUNITY

Matthew W. Wilson

Projects like the Participatory Geographic Information Systems for Transportation (PGIST) (see section by Nyerges, Ramsey, and Wilson, above) engage the public by encouraging them not only to examine the indicators used by regional planning organizations but also to participate directly in the creation of these indicators. Neighborhood-scale decisions in Seattle have involved the public in different ways, by creating opportunities for residents to conduct walking surveys of their community streets and create action plans to address neighborhood concerns. Sustainable Seattle, a nonprofit organization with international recognition for regional sustainability indicators, partnered with a foundation in New York City to implement a neighborhood sustainability indicator program here. This section is an overview of how participatory mapping allowed residents to identify local problems, and how these "geocoding" practices (entering locational data and type of feature into a database) slipped into forms of urban social control (see section by Herbert above).

From 2004 to 2007, Sustainable Seattle conducted over twenty-five participatory street surveys in a program called the "Sustainable Urban Neighborhoods Initiative" (SUNI). Participants collected geographic data about community "deficits" and "assets," using hand-held devices

TABLE 5.3 TOP FEATURES AND CONDITIONS CODED BY SUNI

FEATURES

1 sidewalk
2 building
3 tree pit
4 planting strip
5 business
6 roadway
7 crosswalk/ intersection
8 streetlight
9 tree
10 curb
11 trashcan
12 other
13 suspicious activity
14 curb ramp
15 publication/distribution

CONDITIONS

1 graffiti
2 other
3 uneven pavement
4 litter
5 weeds
6 trip hazard
7 invasive plants
8 damaged
9 trip hazard/uneven pavement*
10 pothole
11 local/non franchise
12 flowering/greenery
13 missing
14 fliers/stickers posted
15 clean

*Because of the evolution of SUNI's terminology, this category was later divided into separate conditions (numbers 3 and 6).

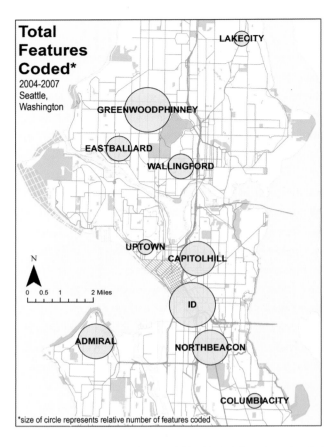

5.16 Proportional symbol map relating the total number of coded *feature-conditions* in SUNI by Seattle neighborhood.

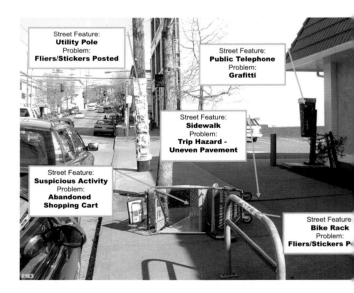

5.17 Annotated photographs were used to train Seattle residents how to read their city streets as *feature-conditions*.

while walking around their neighborhoods. These residents geocoded graffiti, litter, vacant buildings, abandoned automobiles, and sidewalk obstructions, as well as "friendly" business districts, appropriate building facades, peopled sidewalks, and healthy vegetation.

Participants were actively identifying detriments and attributes of their neighborhoods, but their geocoding practices also demonstrated how technology, urban revitalization, government-performance measurement, and quality-of-life indicators are increasingly inextricable. Public and private spaces are now being linked to expansive data networks through sophisticated mobile and wireless geographic information technologies. The city in which we live is both "out there" in the real world and "inside" GIS.

The SUNI survey recorded nearly 6,000 entries over the course of the initiative, where each entry is a mapped location in one of the ten neighborhoods surveyed. These entries were either deficits or assets, and each deficit or asset was further characterized by the feature being coded and the condition of that feature. In total, 68 different features were coded in the street surveys, and 178 different conditions were used to describe those features, reflecting the ways in which the terminologies became more numerous as the survey was applied in different neighborhoods, over time. Table 5.3 describes the top fifteen features and conditions coded in the surveys, and figure 5.16 depicts the proportions of this survey work across the ten neighborhoods coded in Seattle.

Drawing on a narrative frequently considered by participatory GIS proponents, Sustainable Seattle maintained that these participatory mapping practices empowered residents to make their neighborhoods more livable and, thereby, more sustainable. These technologies provide a way for local organizations to support and improve their work, and a way for residents to get involved. They link citizen empowerment with indicators of governmental accountability (Did problems get solved?) and neighborhood quality of life (Has the neighborhood improved?).

Yet, these "feature-conditions" also exemplified how residents' geocoding practices made certain aspects of urban living meaningful—with particular attention to how feature-conditions like graffiti and "suspicious activity" motivated certain imaginations of urban bodies. Figure 5.17 is one slide used to train residents to read their neighborhood streets as feature-conditions. Here, the distinctions among the citizens, the state, technology, and data become fuzzy, as each helps define (or constitute) the other. Data collected through the use of these hand-held technologies make the neighborhood resident into a geocoding subject, who through the technology and software is programmed to visualize the street as a series of discrete, knowable objects, whether they are or not.

After completing their training, residents would walk specified streets in their neighborhood to geocode features like sidewalk cracks and broken street signs. The coding of these features of the built environment enables a neighborhood to alert their government to things needing repair. The features also indirectly identify certain urban bodies as deficits needing governmental attention, namely deviant youth and the homeless. Residents walking their streets would geocode graffiti tags and the presence of shopping carts or alcoholic beverage containers in an effort to map areas of decline within their neighborhood. In this way, geocoding surreptitiously enables the easy construction of a causal relationship between the disorder of the built environment and social disorder—further extending the broken-windows worries within a geocoded urban political geography.

NONPROFIT ORGANIZATIONS AND THE URBAN POLITICAL GEOGRAPHIES OF SEATTLE

Sarah Elwood

Nonprofit and voluntary organizations in Seattle have played a long and significant role in shaping the city's urban and political geographies.[14] Political geographers have a long-standing interest in the ways in which individuals and organizations come together to create and modify their communities, and especially in the changing role and impacts of nonprofit organizations, voluntary groups, and other purportedly grassroots initiatives that work in the city. This section provides examples of some of the ways in which nonprofits shape the geographies of Seattle, and discusses some of the challenges they encounter in doing so. Examples are drawn from a group of thirty organizations that have been partners in a community-engaged, undergraduate geography class at the University of Washington from 2007 to 2009. This course, "GIS Workshop," helps Seattle's nonprofits build access to geographic information systems (GIS), a computer software used for data analysis and mapping. In the workshop, small teams of students collaborate with local nonprofit organizations to carry out a GIS analysis and mapping project for that organization over the course of an academic term.

GIS is a computer system widely used in a diverse range of activities, including environmental management and social-service delivery. For example, once information, such as census data on housing, demographics, income characteristics, or information showing the locations and services offered by city food banks is stored in GIS, it can be analyzed and mapped. With these data, the GIS user might examine whether the food banks were located in areas of likely need, as inferred from the census data. GIS was not originally created with nonprofit organizations in mind as users. The cost of the hardware and software and the specialized skills needed to use them remain a barrier, yet a growing number of nonprofit groups have adopted this technology.

There is a great deal of variation in the characteristics of nonprofit organizations. Institutions such as United Way have multimillion dollar budgets and provide thousands of programs and services nationwide. At the same time, some small neighborhood groups with no paid staff and little budget may also be considered nonprofit organizations, relying primarily on the voluntary efforts of their members to carry out activities in their local areas. From the tiniest groups to the largest, the nonprofit sector has experienced tremendous change in the past twenty years. Local, state, and federal governments have scaled

back their involvement in urban planning, economic and community development, provision of social services, and a host of other arenas, in many cases contracting with nonprofit organizations to carry out some of these activities and services instead. These shifts have been accompanied by tremendous growth and professionalization of the nonprofit sector. Between 1998 and 2008, the overall number of nonprofit organizations in the United States grew by 30 percent, such that in 2008 over 1.48 million groups were recorded by the federal government as nonprofits. During this same period, a professionalization in the nonprofit sector has meant growing use of information technologies, emergence of specialized degree programs for nonprofit professionals, and even specialized consultant services for nonprofit groups. Amidst all of this growth, the sector still faces many challenges. While the responsibilities of nonprofits have grown, budgets and staff capacities have remained steady or even declined. More organizations now compete for the same public and philanthropic grants. The economic downturn of 2008–10 has only worsened these challenges.

The experience of nonprofit organizations in Seattle in many ways mirrors these nationwide trends, with the past two decades showing tremendous expansion in both the presence of the organizations in the city and the range of activities in which they are involved. Just as the economies of Seattle and its surrounding area dominate the state of Washington, Seattle-based organizations dominate the state's nonprofit sector, accounting for 70 percent of nonprofits in the state and generating 95 percent of nonprofit revenues. In addition to the presence of very large high-profile philanthropists, such as the Gates and Allen families, Seattle has long shown a relatively high level of individual philanthropic giving compared to other major cities, ranking tenth among the nation's top fifty cities for charitable giving in the mid-1990s.

Seattle's universities are quite active in supporting nonprofits in the region, providing research, technical assistance, and student service-learning partnerships through such centers as the University of Washington's

Nancy Bell Evans Center on Nonprofits and Philanthropy and the Carlson Leadership and Public Service Center; Seattle University's Center for Service and Community Engagement and Center for Nonprofit and Social Enterprise Management; and Antioch University's Center for Creative Change. Finally, the city of Seattle has long supported the involvement of nonprofit groups, voluntary organizations, and residents in local planning, politics, and problem solving. Founded more than twenty years ago, the city's Department of Neighborhoods and its networks of district councils facilitate a great deal of these interactions. Neighborhood-based nonprofits are among the most active participants in Seattle's district councils, and their work is frequently supported by matching grants from the Department of Neighborhoods.

Political geographers have long focused on spatial inequalities of all kinds, and in urban contexts, a key focus has been the uneven distribution of resources and opportunities in cities and the disproportionate impact of these inequalities upon marginalized individuals and social groups. Public transportation may not adequately serve areas where jobs are most available to low-income workers. Affordable housing may be located far from jobs, or concentrated in areas of the city that are devoid of everyday health care, food, transit, and other resources that residents need. Affordable, healthy food may be readily available in some neighborhoods and practically inaccessible in others. Seattle nonprofits do a great deal of work to identify such spatial inequities and assist the individuals and communities affected by them, as illustrated by the work of an organization called Solid Ground.

Solid Ground was founded as the Fremont Public Association in 1974, during a time of recession and growing poverty and unemployment in Seattle. It originally sought to address the needs of people in the Fremont neighborhood, then one of the city's poorest, through basic direct-service programs such as emergency food banks and employment counseling. Today Solid Ground operates over thirty programs throughout the city, providing housing assistance and other programs to alleviate homelessness; transportation and mobility programs to help low-income individuals, seniors, and people living with disabilities; and food and food-security programs to fight hunger. Solid Ground also combats racism and promotes community-building activities.

Of particular note are Solid Ground's transportation, food security, and youth philanthropy programs. Its Working Wheels program responds to Seattle's uneven housing, employment, and transportation geographies. Living-wage jobs are increasingly not located in areas where low-income residents live, and limited availability or the complete absence of public transportation renders many jobs inaccessible to these residents. Working Wheels provides used cars to individuals whose employment is at risk because of transportation difficulties, and also provides low-cost auto repair services. Another well-known Solid Ground program is its Community Fruit Tree Harvest. Through this program, volunteers harvest plums, pears, apples, and other edible fruit from trees made available to the program by private homeowners. The fruit from each year's harvest, usually around 14,000 pounds, is delivered to food banks, meal programs, senior centers, and shelters that distribute the produce to low-income individuals. Part of Solid Ground's hunger and food-security programming, the Community Fruit Tree Harvest illustrates what an important role nonprofit organizations and voluntary programs play in tapping previously unrecognized resources for local needs.

Solid Ground's Penny Harvest program involves Seattle youth in the city's active philanthropic efforts. Each year, children at approximately fifty Seattle area schools organize fundraising drives in which they collect loose coins, later forming councils that deliberate on how these funds will be used. Most schools choose to make small grants to nonprofit organizations or to fund service projects that the children can carry out in the community. In some instances, the children support organizations working to meet needs in their own communities, as with John Hay Elementary School's 2009 grant to the Queen Anne Helpline, which provides housing and food assistance in the school's community. Other children choose to support

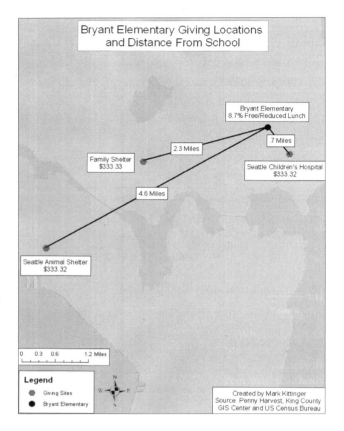

5.18 2009 Penny Harvest grants by Bryant Elementary School.

programs outside their local areas whose services are city-wide, as in the case of Bryant Elementary School's 2009 grants to Family Shelter, Seattle Children's Hospital, and the Seattle Animal Shelter.

Figure 5.18 is one map from a larger research project undertaken for Penny Harvest by the 2009 GIS Workshop. In this project, Penny Harvest sought to better understand the social and spatial characteristics of children's philanthropy, using their Seattle schools as a case study. With Seattle School District data on student demographics and Penny Harvest data on the grants made by each school, the GIS Workshop students examined where schools made grants and what issues they were targeting with their grants, looking for differences in the grant-making choices of the participating schools. Penny Harvest has used the project findings to inform ongoing revisions to their pro-

gram's recruitment strategies and student activities, as well as in their grant-writing efforts.

Another important contribution of nonprofit and voluntary organizations to the urban social geographies of Seattle is their effort to create and preserve open space, community gardens, and public parks throughout the city. These organizations also work to ensure that such spaces are accessible to disadvantaged residents. Seattle's P-Patches provide a place for urban dwellers to grow their own food, make available much-needed green space in urban neighborhoods that often lack parks, and transform previously vacant plots of land into shared spaces that are collectively maintained by the residents who garden in them (fig. 5.19). Nonprofit groups, such as the P-Patch Trust, play a critical role in the complicated process of securing and sustaining these uses of urban lands, and of ensuring their accessibility to all. The P-Patch Trust works with Seattle policy makers and municipal service divisions to achieve long-term tenancy for gardens on publicly owned land and access to water for irrigation; raises funds to waive garden plot rental fees for low-income individuals; and coordinates with other nonprofit organizations to distribute extra produce to food-security programs around the city. This program also illustrates the partnerships through which many nonprofit organizations carry out their work. The P-Patch program, for example, partners with Solid Ground to facilitate the distribution of extra produce to local food shelves and shelter programs.

Seattle's P-Patch gardening program provides an example of nonprofit efforts to ensure public green spaces, even within dense urban areas where land availability and costs are at a premium. The program began during Seattle's economic recession of the 1970s, through the efforts of a University of Washington student, a family-owned truck farm, Seattle's public schools, and local gardeners. Today the P-Patch program involves sixty-eight community gardens, where Seattle residents can pay a nominal administrative fee for a small plot to grow edible and decorative plants (fig. 5.20).

In 2009 the GIS Workshop became involved with the

P-Patch program through the University of Washington's Green Futures Lab. The Green Futures Lab is a community-engaged research group that focuses on sustainable urban design and has collaborated with the P-Patch program in recent years. The students produced a series of maps intended to identify feasible sites for new gardens and plan programs aimed at making urban gardening more accessible to Seattle residents who are underrepresented in the program, including low-income families, immigrants, and racial and ethnic minorities.

While the work of nonprofit and voluntary organizations meets critical needs, over the past twenty or more years, governments have devoted less and less money to supporting nonprofit organizations. During this downturn in funding, many nonprofits have also taken on expanded responsibilities, as government agencies have "devolved" many services onto the nonprofit sector. Amid growing competition for dwindling funds, public and philanthropic funders increasingly expect nonprofit organizations to be able to document the material impacts of their work, creating new imperatives for many nonprofit groups to focus their efforts on activities that will generate demonstrable outcomes, such as the creation of jobs, generation of revenues, or building of new homes.

Nonprofits in Seattle, as elsewhere, have developed a variety of creative responses to these challenges, building on their deep experience of trying to maximize the impact of limited resources. One strategy that has proven fruitful for the organizations discussed above is forming dense networks of relationships with other institutions and organizations, in order to gain assistance from a wide range of sources, exchange and share resources, and advocate for one another. These relationships may bring together local government officials, private businessmen, nonprofit officials, university students and faculty, and local residents.

Seattle's P-Patch Gardens are sustained by just such a mix of diverse partners. The P-Patch Trust provides advocacy and outreach, while the P-Patch program is administered day to day by the city's Department of Neighborhoods. University of Washington initiatives have

5.19 Belltown P-Patch, summer 2009.

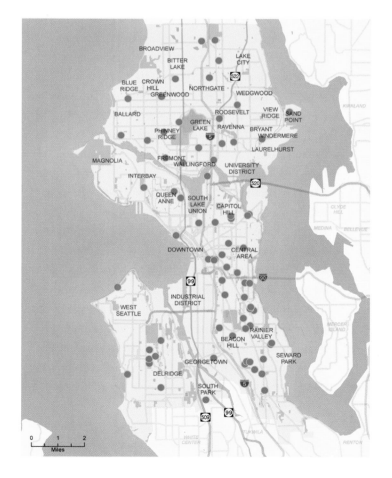

5.20 Seattle P-Patch gardens.

undertaken research to inform planning and outreach for the program. Partnering with other nonprofits also enables the P-Patch program to rely on the already existing knowledge and networks of these other groups, as in their effort to share extra vegetables through Solid Ground's existing network of volunteers and programs for distributing food to families in need. Through these relationships, formal and informal, many nonprofits are able to cobble together a collage of resources, advocates, and institutions that can accomplish more than what would be possible operating on their own.

Another significant development in the work of nonprofits is their increasing reliance on information technologies as part of their work. As nonprofits have expanded their responsibilities and been pressured to document their impact and outcomes, these groups have turned to the Internet, digital databases, geographic information systems (GIS), and, most recently, social networking and interactive mapping technologies, such as Facebook or GoogleMaps. Given the place-based nature of the work that many nonprofits do, GIS has proved especially popular for groups in Seattle and elsewhere. Nonprofit groups use spatial data and maps for research and outreach. One map, for example, was prepared for the P-Patch program, as part of a project aimed at identifying areas of the city underserved by the existing P-Patch network, as well as to recruit potential gardeners to work in existing city gardens. Solid Ground's Penny Harvest program has done extensive analysis of the giving patterns of schools participating in their program, as shown in figure 5.18, a map showing the location and programs of Bryant Elementary School's 2009 grants. These efforts to map the giving patterns of Penny Harvest schools provide one way for program staff to demonstrate the impact of their program in the city, but also give the staff a way to understand key questions about youth philanthropy.

Seattle's nonprofit sector, like that of city's around the United States and worldwide, plays a major role in providing much-needed services to disadvantaged groups,

coordinating and sustaining philanthropic efforts, and advocating for a wide range of social justice concerns. Much of their work is intensely geographical and political, focused on the gaps, injustices, and inequalities created by uneven development across cities and regions. Presently there are new pressures of increasing competition for funds, vastly expanded programs and responsibilities, and newly complex working relationships with public, private, and other nonprofit institutions. Given the stated (and funded) commitment of the Obama Administration to support and help expand the role of nonprofit organizations, as well as Seattle's own robust history of nonprofit engagement, it is clear that these groups will continue to play an active and vital role in the life of the city.

NOTES

1   *United States v. Washington*, 384 F. Supp. 312 (W.D. Wash., 1974).

2   M. Larimer, D. Malone, M. Garner, D. Atkins, B. Burlingham, H. Lonczak, K. Tanzer, S. Clifasefi, W. Hobson, and G. Marlatt, "Health Care and Public Service Use and Costs Before and After Provision of Housing for Chronically Homeless Persons with Severe Alcohol Problems," *JAMA* 301 (2009): 1349–57.

3   Katherine Beckett and Steve Herbert, *Banished* (Oxford: Oxford University Press, 2009).

4   James Wright and Julie Lam, "Homelessness and the Low-Income Housing Supply," *Social Policy* 17 (1987): 48–53.

5   Timothy Gibson, *Securing the Spectacular City* (Lanham, MD: Lexington Books, 2004).

6   James Q. Wilson and George F. Kelling, "The Police and Neighborhood Safety," *Atlantic Monthly* (May 1982): 28–38; George Kelling and Catherine Coles, *Fixing Broken Windows* (New York: Free Press, 1998).

7   Bernard Harcourt, *The Illusion of Order* (Cambridge, MA: Harvard University Press, 2001); Andrew Karmen, *New York Murder Mystery* (New York: New York Uni-

versity Press, 2006); Ralph Taylor, *Breaking Away from Broken Windows* (Boulder, CO: Westview Press, 2001).

8   Steve Herbert, *Citizens, Cops, and Power* (Chicago: University of Chicago Press, 2006).

9   Steve Herbert and Katherine Beckett, "This Is Home to Us," *Social and Cultural Geography* 11 (2010): 231–45.

10   Ibid.

11   Transitioning refers to skating on curved surfaces that transition to vertical walls, such as the sides of an empty pool or concrete drainage ditch. Transition skating was the most popular and publicized skateboarding discipline from the late 1970s through the late 1980s.

12   The authors contributed to this section equally. We acknowledge co-researchers who greatly contributed to the design and development of the Let's Improve Transportation Web site: Adam Hindman, Jordan Isip, Mike Lowry, Martin Swobodzinski, Zhong Wang, and Guirong Zhou. We also thank Robert Aguirre for his assistance in providing some preliminary analyses of the user-event logs.

13   At the time of publication, the LIT Web site is available to view online. The site is secure and can be safely accessed, but note that a security warning may appear before you can access it.

14   I am grateful for material contributed to this section by Jessica Nguyen, Christopher Slotta, students and teaching assistants from my Spring 2007, 2008, and 2009 GIS Workshop classes, and our community partners in these workshops.

SIX

SOCIAL GEOGRAPHIES

Richard Morrill, Suzanne Withers, Tricia Ruiz,
Mark Ellis, Catherine Veninga, Kim England,
Gary Simonson, Tony Sparks, Michael Brown,
Sean Wang, and Larry Knopp

Social geography explores the relationship among groups, identities, and places. Who lives where in Seattle, and why? How does housing and education vary across neighborhoods in the metropolitan area? How do groups use parts of the city to assert their identity or to contest others' identity? We begin with a population-geography of the area, showing where different sorts of people live, and then move on to consider the current housing crisis, specifically through an analysis of the local mortgage crisis and through an ethnography of a homeless camp in Seattle. We explore the question of schools and segregation, from both current and historical perspectives. Finally, we offer a series of neighborhood analyses, showing the causes and consequences of important social changes in places like Belltown, Columbia City, and Capitol Hill, among others. These three areas are featured because they exemplify key social processes of race, class, and gender dynamics that are related to broader economic, political, and cultural changes in society generally.

THE PEOPLE OF SEATTLE

Richard Morrill

Figure 6.1 maps the growth of Pugetopolis from 1950 to 2000. The Seattle urban area has grown from 789,000 in 1950 to 3,148,000 in 2000, a growth of 300 percent. Since 1940, growth of the region has averaged a rather high 20 percent per decade, about twice the national rate. Thus the population doubled from 750,000 to 1,500,000 from 1940 to 1965, and doubled again, from 1.5 to 3.0 million, from 1965 to 2000. Some economic bases waxed and later waned, some new economic bases of growth appeared, and, as citizens know well, growth has been volatile, typically with

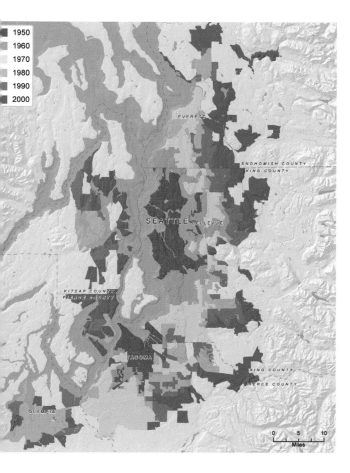

6.1 Map of "Pugetopolis," showing urbanized areas by decade, 1950–2000.

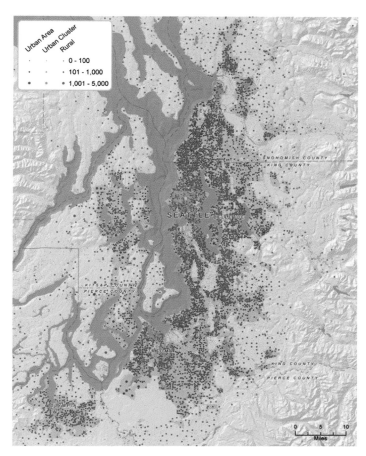

6.2 Urban vs. rural in the year 2000.

recessions early in each decade, followed by late decade booms (except, of course, the present 2000s decade).

The process of urban development has been remarkably consistent: a combination of "up and out," or new suburban growth simultaneous with higher urban density. These outcomes reflect the differing household preferences as well as the location of employment (see chapter 2). In 1950 the urban area included the cities of Seattle and Tacoma, Lakewood, most of Shoreline and Renton, and part of Highline, with a toehold to the east on Mercer Island, Kirkland, and Bellevue. The urban boom of the 1950s extended the urban area up to Edmonds, Lynnwood, and Bothell, east through Bellevue, and south to the rest of Highline in to Kent; Tacoma added more of Lakewood and Parkland. By 1970 the urbanized areas had merged, cur-

rently extending from Fort Lewis all the way through Everett. By 1980 two new, separate, urbanized areas appeared: Bremerton and Olympia, with the urban core now extending to Spanaway, Gig Harbor and Bonney Lake, Soos Creek, Mill Creek, and Marysville. In the 1980s suburban expansion continued: Bremerton added Silverdale and Port Orchard, Tacoma spread southeast, and the Seattle imprint spread almost to Maple Valley, adding the Sammamish plateau and Woodinville, North Creek, and Silver Firs.

Critics view these fifty years of urban expansion as classic "urban sprawl," but this is not completely true. Growth has been mainly the result of the large population increase in the area. Indeed, the average density (which did decline from 1950 to 1970, in the postwar suburban boom) has risen over the last thirty years. However, figure 6.2 shows the

TABLE 6.1 "PUGETOPOLIS" POPULATION GROWTH, 1950-2000

POPULATION (IN THOUSANDS)

	1950	1960	1970	1980	1990	2000
Seattle	622	864	1238	1392	1744	2181
Tacoma	168	215	233	402	497	646
Bremerton				64	112	178
Olympia				69	95	144
TOTAL	879	1079	1571	1927	2448	3148

distribution by block groups of the urban and remaining rural population for the year 2000. The intermingling of red (urban) and green (rural) starkly denotes the zone of sprawl.

Growth Management

Overall settlement patterns in the Seattle region since the mid-1990s have been guided by the Washington State Growth Management Act. In this region, the Puget Sound Regional Council acts as the coordinating body for plans spanning the counties and cities located in the region. It effectively plans for the system of nodes and transport routes. Figure 6.3 shows an overview of this plan, which allows for a hierarchy of settlements, ranging from metropolitan cities to small cities. It also designates regional growth and manufacturing centers, key components of the transportation network, and the urban growth boundary, within which is to be located most new housing and economic activity.

In recent years, only 4 percent of new housing units have been located on rural lands (both inside and outside the urban growth boundary). The Growth Management Act anticipates that unincorporated areas inside the boundary will become parts of cities, and these cities, rather than the county governments, will provide urban services. Lands outside the urban growth boundary are to be managed for their resource values, including agri-

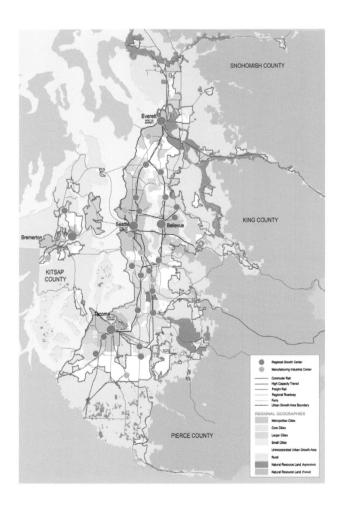

6.3 *Vision 2040* regional growth strategy. Reprinted with permission from the Puget Sound Regional Council.

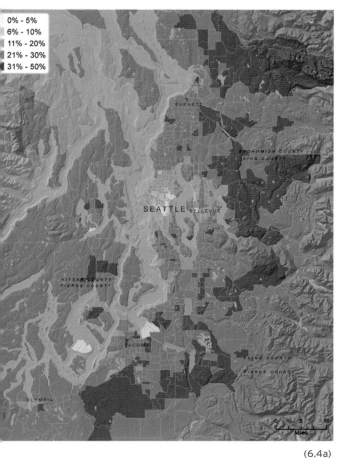

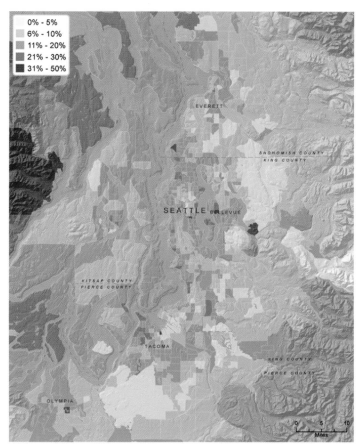

(6.4a) (6.4b)

6.4 The young and old population in Seattle, 2000:
(a) children under ten; (b) the elderly. Source: U.S. Census.

culture and forest products. The current urban realm is
brushing up against the boundary. Growth management
did succeed in reducing the pace and pattern of growth at
the edge, and, in particular, it reduced growth in designated rural areas beyond the boundary.

Demographics

Demography makes much of sex and age, as these under-
lie fertility and survival. Across the urban landscape,
variation in these factors is perhaps not as great as many
might expect. Males and females are in balance over most

of suburbia, but males outnumber females on military
bases, in prisons, around Skid Road areas, and in the
farthest rural reaches—leftovers from the logging era.
Women and the elderly outnumber men in areas with
large nursing homes and retirement communities and,
more generally, in older, inner-city areas. Children tend to
be found in suburbs, but also in areas with high numbers
of recent immigrant families. Older persons are common
in inner suburbs, where people have aged in place (fig.
6.4). Young adults dominate around universities. The
traditional nuclear family represents 40 percent of households in the older suburbs close to urban centers, and
typically less than 20 percent in inner city areas (fig. 6.5).
Seattle's nuclear family count is remarkably low, even for
central cities in the United States. This structure is primarily one of preferences, not of economics.

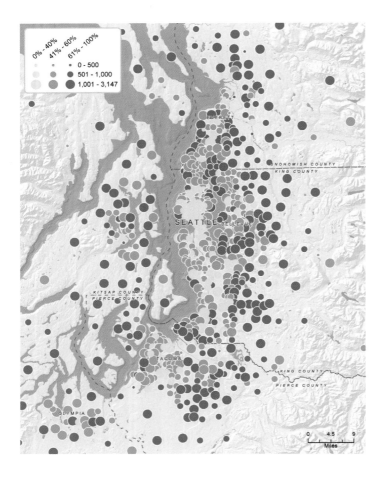

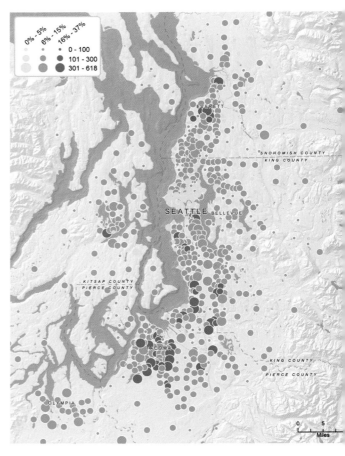

6.5 Traditional nuclear families, 2000. Source: U.S. Census.

6.6 Single-parent families, 2000. Source: U.S. Census

The Census defines heterosexual-childless families in three ways. Most are empty-nester families, where grown children have left home. Some are newly formed families, with children yet to come, and some couples are childless on purpose (but classed as families by the census/society, simply because they are "married"). These categories have an interesting geography. Over 40 percent tend to be in older, richer suburbs (or city areas), such as in Mercer Island and Bellevue, probably mainly empty nesters. Quite high percentages (35–40 percent) occur in very low-density suburbia, probably a consequence of purposeful migration. Again, surprisingly low percentages prevail in most of Seattle, except for the richest waterfront tracts.

In the Seattle region, single-parent families (fig. 6.6), about 85 percent headed by females, are strongly associ-

ated with poverty and with black and Hispanic populations. Percentages are highest in southeast Seattle and in south King County and south Everett, in areas with subsidized public housing.

"Non-family households with two or more persons" is the Census Bureau's terminology for roommates and partners, opposite sex and same sex (fig. 6.7). As many as half are de-facto families in the sense that many husband-wife families without children are. Their prevalence is unusually high in Seattle, and they are the most rapidly increasing type of household. Unmarried, opposite-sex partners are especially prevalent from north Capitol Hill in Seattle through Queen Anne, Fremont, Ballard, and Greenwood, a young population closely tied to the University of Washington and high-tech and medical job

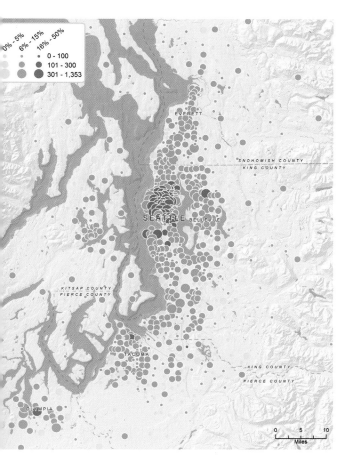

6.7 "Non-family households with two or more persons," 2000. Source: U.S. Census

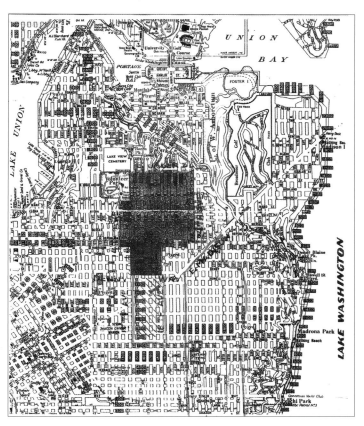

6.8 Restrictive covenants in Seattle, ca. 1950–1960. The green line encircles the part of Capitol Hill that was signed and covenanted. The pink line indicates an area where some people hoped to achieve the same end through "friendly cooperation." Source: From the collection of Richard Morrill. Original source unknown, though probably made in the 1950s by a real estate entity long since gone.

opportunities around Lake Union. Other concentrations are in the SR-99 corridor in and south of Everett, in quite a few areas of Tacoma, and in Auburn, Kent, and Federal Way, most in less-affluent, apartment-living areas. Unmarried, same-sex partners are discussed later in this chapter. Single-person households are concentrated in apartment districts (e.g., in downtown Seattle, Wallingford, and Ballard), and there are a few suburban apartment tracts in downtown Kirkland, Bellevue, Kent, and Edmonds (where they dominate).

Historical Geographies of Race

Far from the American South and Mexico, Seattle was historically a very "white" city, especially low in Hispanic and black populations, but with a higher Asian share, because of the city's Pacific port-of-entry position. Native Americans occupied many village sites around the current metropolis, but by 1880 were relegated to now far suburban reservations. In the 1940s to 1960s, Seattle's black community became a fairly segregated "ghetto," growing in the 1940s and 1950s by migration from Texas and Louisiana, and since the 1960s from California. The racism of the period was exemplified by the restrictive covenant to stop northward expansion of the ghetto (fig. 6.8). Group action and legal decisions led in the late 1960s to the end of formal

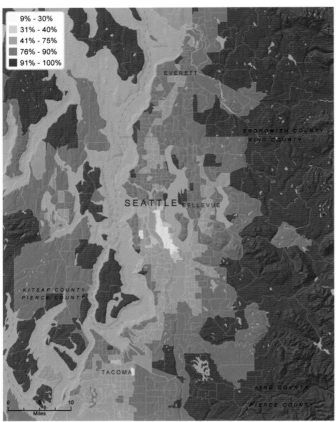

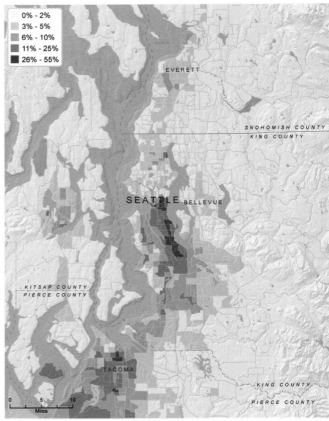

(6.9a)   (6.9b)

discrimination in the real estate market, and school desegregation in the late 1970s was carried out without court
action. Growth to the southeast was aided by suburbanization—in part, white flight—of the middle-class populations.
By the 1980s, gentrification of the northern part of the community was leading to displacement southward.

The early Asian population, mainly Chinese, lived in
what is now called the International District, surviving
despite a general Chinese exclusion in 1910. Workers from
Japan were brought in by contract in the 1920s for work in
agriculture, mainly in south King County. They gradually
moved into cities, but were rounded up and sent to internment camps, primarily in Idaho, Montana, and Wyoming,
at the start of World War II, returning to Seattle and Tacoma
only after the war ended. The next wave of Asians were
refugees, first from Korea (1950s), then from Vietnam and
Cambodia (1960s, 1970s), along with slow but steady migration from the Philippines, and then, mainly since 1990,

from China, India, and Pakistan, often recruited or enticed
by high-tech job opportunities, including at Microsoft. The
recent increase in the Hispanic population was unexpected
and seems to some extent a spillover or re-migration from
agricultural parts of the state and by construction opportunities in Seattle.

Current Patterns

Despite its history of being a very white American city,
Seattle is becoming more diverse. Figures 6.9a-e show the
clustering of white, black, Asian, Native American, and
Hispanic populations. Minority groups are obviously clustered, but not very strongly. There are no "ghettoes" in the
historic sense of total segregation. The black population
is the most clustered, the main area extending southward
from its original core east of downtown Seattle (the Central
District or CD), southeasterly down Rainier Valley into sub-

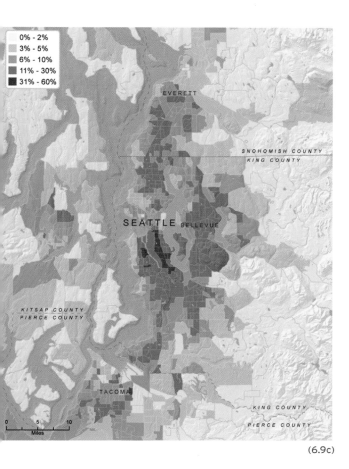

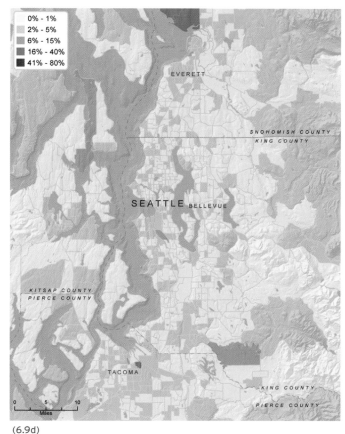

(6.9c)   (6.9d)

6.9 Racial distribution in 2000: (a) white; (b) black;
(c) Asian; (d) Native American; (e) Hispanic. Source: U.S. Census.

urban Bryn Mawr–Skyway and Renton, and with a similar
smaller community west of downtown Tacoma (Hilltop).
In 1990 the degree of segregation was moderate by national
standards, and it has fallen since. The Asian American
population is much larger, now quite diverse and also dispersed. A large cluster still extends from the "International
District," just south of downtown Seattle, southeasterly
down Beacon Hill (thus lying west of black concentrations), and also extending into Renton and farther suburbs.
There are also significant numbers in far north Seattle, into
Shoreline and southern Snohomish County, in parts of Bellevue and Redmond on the Eastside (and including many
recent immigrants), and in cities of south King County,
such as Kent, Federal Way, and Auburn, home of an historic, mainly Japanese farmworker population. Tacoma has
sizeable Asian communities as well.

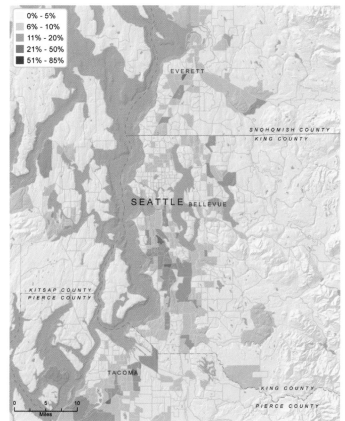

(6.9e)

The Native American population resides predominantly on several reservations, in particular in Kitsap and Snohomish counties, but many are dispersed across the metropolis. The Hispanic population was still low in 2000, but has since almost doubled. The greater concentrations are in south King County, in the towns of White Center, Burien, and Kent, rather than in the city of Seattle.

The percentage of persons who claim mixed racial heritage is high in the Seattle region, but the city of Tacoma stands out with an unusually high share of almost 5 percent, followed by parts of south Seattle and around Bremerton. The percentages have been rising further since 2000, partly from immigration but partly also because more people are willing to recognize their mixed ancestry.

Social Class

Arguably, data in figure 6.10, which shows median household income, are the single most informative descriptors of the social geography of the metropolis. Areas with the highest incomes are the far suburbs of the Eastside, from Maple Valley through Sammamish; Cottage Lake to Mill Creek; waterfront and view areas in Mercer Island, Bellevue, Kirkland, and Bainbridge Island; and from Edmonds to Mukilteo. View and waterfront tracts in Seattle are included as well, but, overall, Seattle ranks less high, because it has so high a share of nonfamily households (these areas would look "higher" on a per-capita income measure). Intermediate incomes prevail over much of north Seattle, Shoreline, Kirkland, Federal Way, the Gig Harbor peninsula, Marysville, and the rural fringes of all the counties. Lower incomes prevail in the older core towns of Everett, Bremerton, parts of Tacoma and Lakewood, and in the Seattle core, despite gentrification.

The Seattle region as a whole is more educated than the nation as a whole, King County is unusually highly educated, and the city of Seattle, with 52 percent of adults with a bachelor's degree or more, is exceptional (figure 6.11 shows the distribution of those with a BA or more). Two kinds of areas have high percentages—first, wealthy subur-

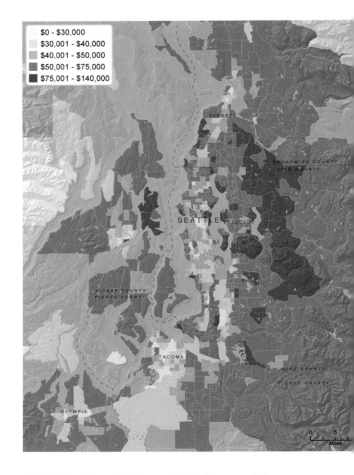

6.10   Median household income, 2000. Source: U.S. Census.

ban areas in general, especially the Eastside communities, and, second, the entire north-central part of the city of Seattle surrounding the University of Washington. Intermediate levels prevail over middle-income sections of the city, away from the university, and in the suburbs, while low levels occur in lower income areas in all four counties, but especially in south Tacoma, Bremerton, and much of south King County, as well as in areas with high percentages of blacks or Hispanics.

Lower class areas include traditional zones of mixed housing, industry, and transport, as in South Seattle, Everett, Bremerton, Auburn, and especially Tacoma. The largest area of lower class neighborhoods extends from South Seattle through south King County to Tacoma,

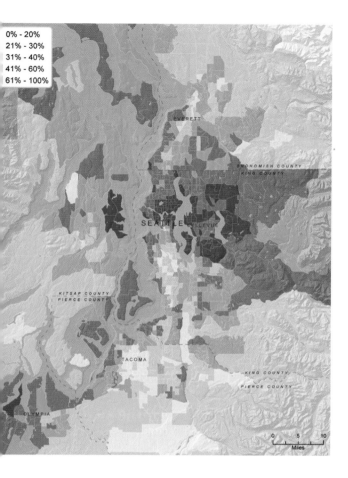

0% - 20%
21% - 30%
31% - 40%
41% - 60%
61% - 100%

6.11 College-educated population, 2000. Source: U.S. Census.

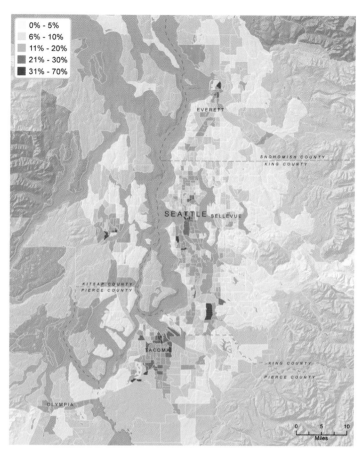

0% - 5%
6% - 10%
11% - 20%
21% - 30%
31% - 70%

6.12 Poverty, 2000. Source: U.S. Census.

marked by historical development, displacement from Seattle, and high minority populations. The second large zone of lower class settlement is the rural fringe, especially in Pierce and Snohomish counties, and may surprise those who think that rural areas are the home of rich estates. The middle class areas (orange and rust) are intermediate in location and dominate the outer suburban areas as well as some older inner neighborhoods of Seattle and Tacoma.

Figure 6.12 maps the percentages of persons below the poverty line. The geography it reveals is similar to but subtly different from the income map (fig. 6.10). High concentrations of poor households occur around the area's universities, but this cluster is of a temporary nature. High-

est poverty occurs in the general International District area south of downtown, in areas with major public housing projects, and around Fort Lewis and the Bremerton Naval Shipyards. Moderately high levels occur in most areas that have high percentages of blacks and Hispanics, or, in south Tacoma and south King County generally. Poverty is low, as expected in the areas of high median income.

Median house values, shown in figure 6.13, have a geography similar to that of median household income, but with an interesting difference. Values in the city of Seattle, and especially in view and waterfront locations and areas near the University of Washington, are very high; and, relatively, homes in south King and Pierce counties are more affordable and have attracted many families who were priced out of the Seattle housing market. Median rents show a fairly different pattern. The highest values are in Eastside cities, especially Kirkland and Redmond, serving Microsoft. Values in Seattle are intermediate (but may impose as high a burden on less affluent renters). Rents are markedly lower in Everett, Bremerton, and Tacoma, where, again, the lower rents may exact a fairly high share of income.

Now that we have sketched out the basic demographic patterns and structure of the region, we turn next to some of the substantive topics in social geography that relate to these basic patterns. We'll start with research on housing.

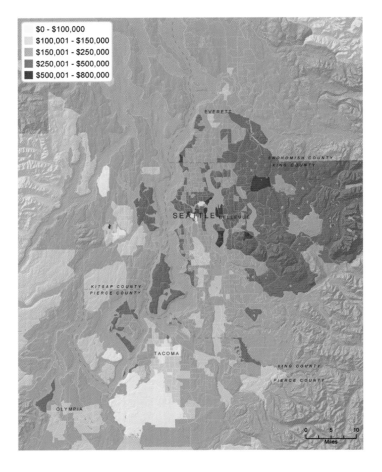

6.13 Median housing value, 2000. Source: U.S. Census.

RIDING OUT THE STORM:
VULNERABILITY IN SEATTLE'S HOUSING MARKET

Suzanne Davies Withers

Although Seattle pales in comparison to places like San Francisco and Washington, D.C., in terms of housing affordability, problems increased there during the prolonged housing boom, particularly from 1995 to 2005. However, the recent downturn in the global economy has fueled significant housing market shifts across the United States. Staggering levels of mortgage delinquencies, foreclosure rates, abandoned properties, and negative equity have plagued community after community across the nation, particularly throughout the states of Florida, California, and Nevada, which were hit hardest by the puncture of the housing bubble. The impact of the housing crisis varies enormously geographically. The following vignette describes how Seattle has been riding out the storm.

Housing Price Trends

Seattle is one of twenty cities included in the Case-Shiller Index of Housing Prices, which serves as the industry standard for measuring house-price changes over time. The index tracks the value of housing as an investment. It is based on the sale prices of standard existing houses, rather than new construction. The benchmark is usually 1990 (value = 100), and the index factors out the effects of inflation. From 1995 to 2005 the equity in housing rose at an unprecedented rate.

Figure 6.14 shows the change in housing prices in a few select cities since 1990 (the benchmark year in this figure is 2000). Compared with other cities, Seattle has had a relatively gradual appreciation in home prices. It has also had a much more gradual depreciation. Seattle most certainly rode the wave up and then back down, but the surge was quite tame compared to San Francisco, Las Vegas, or Miami. This figure also suggests that prices have started to

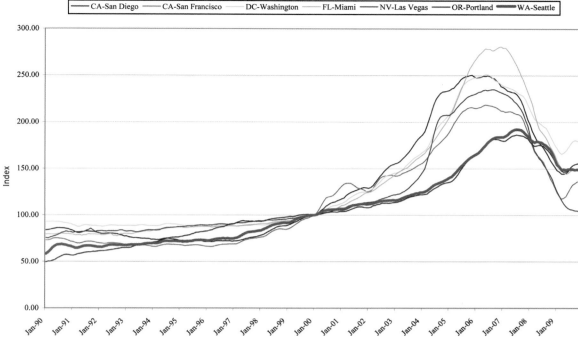

6.14 Change in housing prices since 1990.

appreciate recently, but it remains to be seen whether this momentum will be sustained. Still, compared to much of the nation, Seattle has had a long run of significant housing price appreciation.

According to figures derived from the Home Price Index (a broad measure of the movement of single-family house prices), over the past twenty years, cities across the nation have had a median home-price appreciation of 115 percent, whereas Seattle has experienced a 243 percent appreciation. Within the last ten years the national median appreciation has been 54 percent, yet in Seattle it was 80 percent. The last five years witnessed a median appreciation across the nation of 17 percent, during a time when Seattle experienced 32 percent. During 2008 the national rate was minus 1.8 percent, but Seattle experienced a change of minus 9.5 percent.[1] Having risen higher, we in Seattle had farther to drop (see fig. 6.15).

Housing Affordability

During the period 1990 to 2008, the pressure to get into the housing market was quite intense for those who aspired

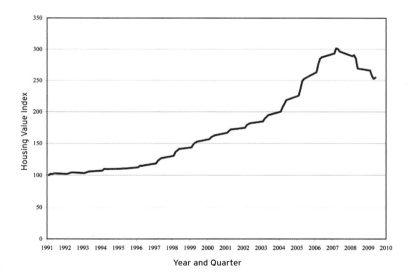

6.15 Home Price Index in the Seattle region, 1991–2009.

to become homeowners. The housing affordability index for Seattle reveals that Seattle is one of the least affordable cities in which to live. In 2000 the median sales price for an existing single-family home was just shy of $250,000 (table 6.2). By the time the housing market peaked in 2007, this value had increased to $455,000. Traditionally, the industry has considered a reasonable ratio between home price and household income to be in the range of 3:1. In 1995 the ratio of median sales price to median household income was 3:7. By 2000 this ratio had increased to 4:5 and steadily rose to 6:7 by 2007. Table 6.2 indicates that while housing prices were rising, incomes were relatively flat.

The housing affordability index measures the ability of the median income to purchase the median priced home, given a scenario of a 20 percent down payment and a traditional thirty-year loan. For all buyers, the index is equal to 100, if there is a balance between median housing costs and median income and given a 25 percent qualifying ratio. A value above 100 indicates that a household has more income than needed to qualify for a typical mortgage loan. A value below 100 indicates insufficient income to buy a typical home. Since 2005, amongst all buyers, the index has been well below 100, hitting a low of 66.4 at the peak of the market in 2007. By 2008, affordability was renewed after the market topped. For first-time buyers, the housing affordability index is based on earners who are 70 percent of median area earners, and housing costs are at 85 percent of the area median price, with a 10 percent down payment. The figures in table 6.2 indicate how unaffordable Seattle housing is for first-time buyers. In 2000 the index was 53.1. During the real peak in the market, the index was as low as

39 in 2006 and 37 in 2007. Now, the post-boom scenario is indicating a return to previous rates of affordability.

Vulnerabilities, Delinquencies, and Foreclosures

As of February 2010, almost 16 percent of homeowners, nearly a quarter of a million households, were under water on their mortgages in Washington State. In the Seattle-Bellevue-Everett metropolitan area the rate was 15 percent. Washington State had only 15 percent of subprime loans, compared with the nation's level of 21 percent. Subprime loans are more likely to default, and, indeed, the subprime adjustable rate loans in Washington had greater increases in delinquency and foreclosures when the bubble burst.

More than two-thirds (69 percent) of the homes for sale in Seattle in March 2010 were foreclosure properties—well over 4,000 homes. What is the geography of delinquencies and foreclosures across King County?[2] To answer this question, we created a data set that merged three data sources: potential foreclosures from January 2005 through August 2009, demographic data from 2009, and property information.[3] Figure 6.16 displays the number of properties at risk of foreclosure. Each red dot represents five properties. King County to the west is completely swamped. Only the very

TABLE 6.2 KING COUNTY HOUSING AFFORDABILITY INDEX

KING COUNTY	2000	2001	2002	2003	2004	2005	2006	2007	2008	2009
Median Sales Price ($)	249,900	262,000	278,500	292,400	324,000	374,000	425,000	455,000	450,000	375,000
Median Household Income ($)	56,100	56,900	58,100	58,800	63,600	62,700	65,800	68,200	68,800	
Median Price/Median Income	4.5	4.6	4.8	5.0	5.1	6.0	6.5	6.7	6.5	
Housing Affordability Index										
All Buyers	92.5	103.2	104.6	121.3	105.1	87.4	70.2	66.4	74.3	102.5
First-time Buyers	53.1	58.7	59.2	68.2	58.4	48.5	39.3	37.2	41.4	57.0

SOURCE: U.S. Census Bureau, Washington State Office of Financial Management, Washington Center for Real Estate Research

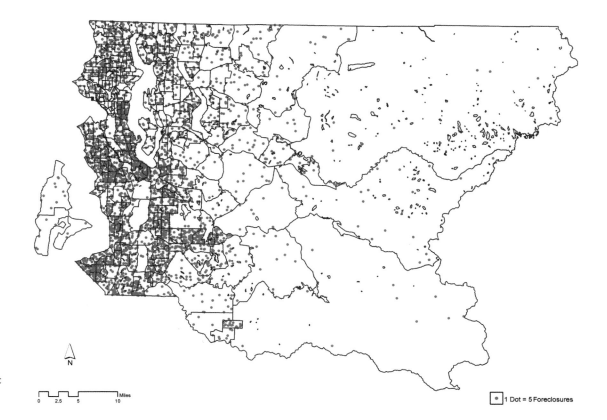

6.16 Properties at
risk of foreclosure.

● 1 Dot = 5 Foreclosures

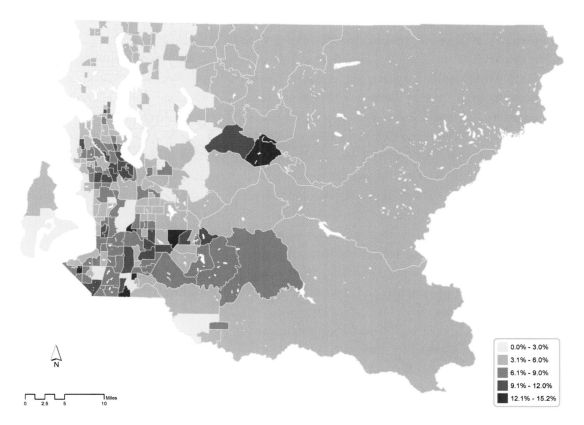

6.17 Rate of fore-
closure filings.

	0.0% - 3.0%
	3.1% - 6.0%
	6.1% - 9.0%
	9.1% - 12.0%
	12.1% - 15.2%

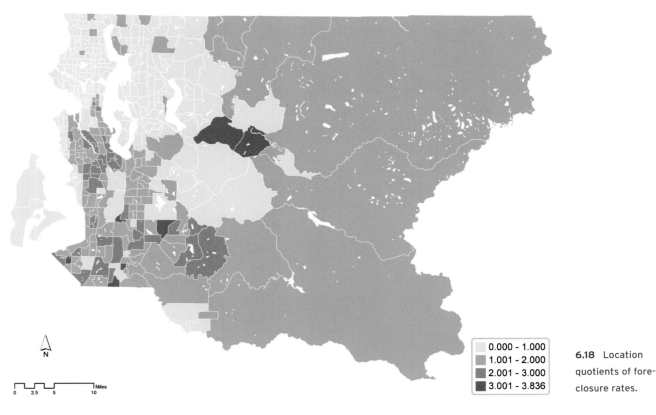

N

0   2.5   5   10   Miles

6.18 Location quotients of foreclosure rates.

large census tracts to the east are not solid red. This variation is due partly to the variation in population density and the number of owner-occupied housing units. A more telling picture is figure 6.17, which reveals the rate of foreclosure filings (calculated by taking the ratio of the number of property filings to the number of owner-occupied homes). The lowest rates are found north of the Ship Canal Bridge, and in the Kirkland and Redmond areas to the east. These areas have a rate of 3 percent or fewer. In contrast many areas to the far east of the county and to the south have rates ranging from 3 to 9 percent. The more extreme rates are geographically concentrated within a few regions of the county. There is a high incidence in the Sammamish-Issaquah area moving toward North Bend. There is also a high incidence in the southern part of the county toward Tukwila and Burien.

Geographers tend to use location quotients to map spatial variation. A value less than 1.0 indicates that a census tract has fewer than its fair share of a variable, given its relative size in the county overall. A value above 1 indicates that a tract has more than its fair share of a vari-

able. A value over 2 indicates that a census tract has twice the expected amount, and so on. Figure 6.18 displays the location quotients of foreclosure risk by census tract. The geographic variation is more pronounced compared to the other maps. Generally speaking, the areas with low location quotients are in the well-established ring cities.

The extremely high rates seem to represent different types of vulnerability. Figure 6.19 displays the proportion of the tracts' populations that are minority groups. There is a weak but statistically significant correspondence between the risk of foreclosure and the proportion minority (r=0.350). There is some correspondence in the Rainier Valley area immediately south of Seattle and down near Federal Way. Otherwise, there is little correspondence. Poverty (see fig 6.12) in a tract does not seem to correspond with the risk of foreclosure (the correlation is weak and negative). Although that may seem counterintuitive, it is likely that areas with high poverty rates will have low levels of homeownership.

Figure 6.20 displays median age of the housing. The correspondence is significant but weak (r=−0.165). A number of the clusters of high foreclosure rates correspond spa-

6.19 Minority population proportions by census tracts, 2000. Source: U.S. Census.

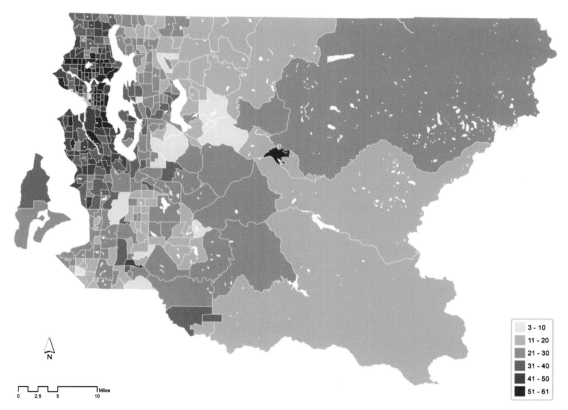

6.20 Median age of housing structure, 2000. Source: U.S. Census.

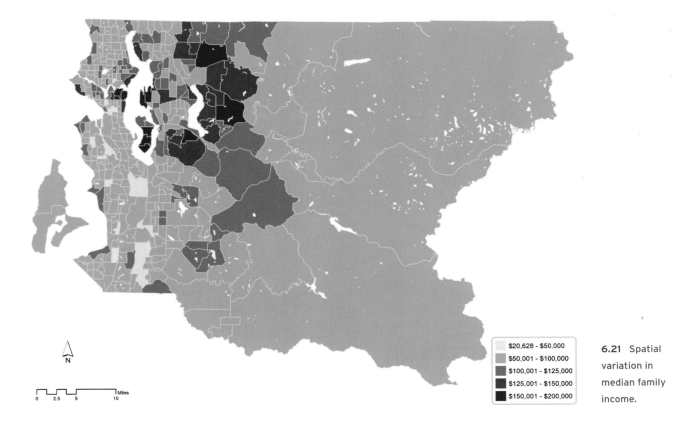

6.21 Spatial variation in median family income.

Legend:
- $20,628 - $50,000
- $50,001 - $100,000
- $100,001 - $125,000
- $125,001 - $150,000
- $150,001 - $200,000

tially to areas where new homes have been built within the past decade. Frequently these are large, expensive homes, at farther distance from Seattle. The risk of foreclosure is high in these areas, to the east especially.

Lastly, figure 6.21 displays the spatial variation in median family income. As expected, it is significant and negatively related to the vulnerability of foreclosures ($r=-0.407$). While it is predictable to notice that some areas have low median incomes and high rates of foreclosure, others have very high rates of foreclosure and high levels of median income. The latter group tends to be in the newly built areas.

Stemming the Tide

The geography of vulnerability, in the sense of mortgage delinquency and foreclosure, varies considerably over space in King County. Like in other parts of the country, the list of who is vulnerable in the economic climate of 2008–10 is mixed. Incomes have been flat, housing prices have been rising, unemployment has been on the rise, and housing consumption levels have reached the point of staggering square footage. Much of this stems from, and contributes to, a severe budget crisis in Washington State.

Seattle and the region have been battered and bruised during this crisis. The vulnerabilities and misfortunes of some, however, represent a new landscape of opportunity for others. The most recent headlines in the Seattle newspaper boast affordability levels previously unimaginable. Interest rates are low and housing is more affordable now than it has been in over a decade. Some economists predict that the intense wave of foreclosures has finally ebbed. In Seattle and King County, it is unlikely that we have seen the end yet, as layoffs continue to occur and foreclosures still make up the lion's share of properties on the market. For those who are employed these are opportune times. Yet, while some people will be in a position to benefit from these housing market shifts, many of us will remain vulnerable for quite some time and many will never be able to regain the ground they lost when the markets fell.[4]

NOTES FROM SEATTLE'S TENT CITY 3:
A SOCIAL GEOGRAPHY
OF A HOMELESS ENCAMPMENT

Tony Sparks

I got off the bus in front of St. Mark's Cathedral at around 4:30 on a late fall afternoon. It was raining lightly and just beginning to get dark. The temperature was hovering around 40 degrees. I had noticed from aboard the bus that you could just see the tops of some tents in an area that was soon to be my new home. Excited to be starting my field research in Seattle's Tent City 3, I shouldered my backpack, crossed the parking lot, and headed to what was normally the side lot of the church, now transformed into a small city. What set this city apart from the larger metropolis surrounding it was that it was comprised entirely of tents. Huge tents, tiny tents, tents made of tarps, military tents, and recreational tents, all set up in a large U shape.

As I approached the front desk of the Executive Committee office, I was a little nervous. It had been a few weeks since the vote was taken that allowed me to become a live-in researcher/resident and I wasn't sure if anyone would recognize me. During that time, the population of the tent city had swelled from about thirty residents to nearly eighty. The person at the desk did recognize me, and he greeted me warmly. I signed the Tent City Code of Conduct (fig. 6.23), after he reviewed it with me. Then he took me on a brief tour of the camp, pointing out the location of communal spaces, includ-

6.22 "Welcome to Tent City 3." Photo by Tony Sparks.

> ## TENT CITY CODE OF CONDUCT
>
> WE, THE PEOPLE OF SHARE/WHEEL, IN ORDER TO KEEP A MORE HARMONIOUS COMMUNITY, ASK THAT YOU OBSERVE THE FOLLOWING CODE OF CONDUCT.
>
> THE SHARE/WHEEL TENT CITY IS A DRUG AND ALCOHOL FREE ZONE. THOSE CAUGHT DRINKING OR USING DRUGS WILL BE ASKED TO LEAVE. SOBRIETY IS REQUIRED.
>
> NO WEAPONS ARE ALLOWED. KNIVES OVER 3 1/2 INCHES MUST BE CHECKED IN.
>
> ANY VIOLENCE WILL NOT BE TOLERATED. PLEASE ATTEMPT TO RESOLVE ANY CONFLICT IN A CREATIVE AND PEACEFUL MANNER.
>
> DEGRADING ETHNIC, RACIAL, SEXIST OR HOMOPHOBIC REMARKS ARE NOT ACCEPTABLE. NO PHYSICAL PUNISHMENT, VERBAL ABUSE, OR INTIMIDATION WILL BE TOLERATED.
>
> WE ARE A COMMUNITY. PLEASE RESPECT THE RIGHTS AND PRIVACY OF YOUR FELLOW CITIZENS.
>
> NO MEN IN THE WOMEN'S TENTS. NO WOMEN IN THE MEN'S TENTS. NO LOITERING OR DISTURBING THE NEIGHBORS. NO TRESPASSING.
>
> ATTENDANCE OF AT LEAST ONE OF THE SEVERAL COMMUNITY MEETINGS HELD THROUGHOUT THE WEEK IS REQUIRED. DAYS AND TIMES WILL BE POSTED SO THAT YOU MAY WORK IT INTO YOU SCHEDULE.
>
> IF THESE RULES ARE NOT RESPECTED AND ENFORCED, TENT CITY MAY BE PERMANENTLY CLOSED.

6.23 Tent City's "Code of Conduct." Reprinted with permission from SHARE/WHEEL.

ing the kitchen and the TV room, as well as the Porta-Potties. As we walked around the camp, he explained that it was divided into sections (see layout in fig. 6.24): large communal tents at the front, private tents around the perimeter, and medium-sized "couples tents" clustered in a group toward the middle (fig. 6.25).

I was home.

Countless accounts of homelessness over the past 150 years tell a similar tale. Although poorhouses no longer exist and vagrancy is no longer a punishable crime, the conditions that create homelessness—the dearth of spaces in which the homeless are allowed to reside and the often condescending attitude of many who want to "help" the homeless—also have remained remarkably consistent. Indeed, in nearly all of the interviews I conducted during my stay in Tent City 3, as well as in countless informal conversations, residents remarked on the assumption, voiced by care workers, policy makers, and the general public, that the condition of homelessness was often the product of some sort of personal failing or pathology. (One

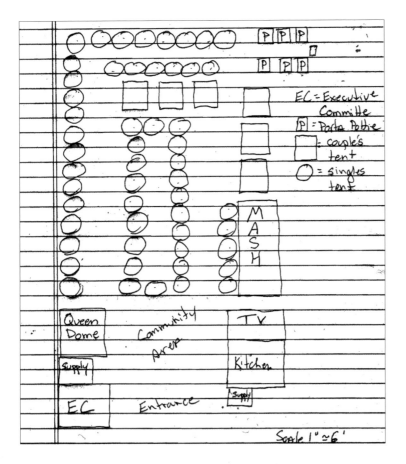

6.24 Layout of the tent city at St. Mark's Cathedral.
Source: From author's fieldnotes, March 25, 2005.

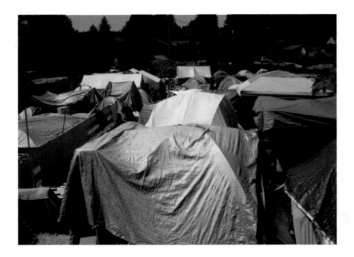

6.25 Tent City 3.

resident sarcastically referred to this so-called pathology as "lazy-crazy-drunk-high it is"). In short, the homeless must, on a daily basis, battle the assumption that because they lack private space, they must be somehow defective; or, conversely, they lack private space because they are defective.

Within this spatial "catch-22," the tent city is more than simply a shelter of last resort. It is a visible symbol of poverty and social inequality marring Seattle's pristine and prosperous landscape—for both proponents and detractors alike. Yet for its residents, it is a space, produced *by them*, to which they can escape from the public's condescending gaze and begin to reclaim their identity, agency, and autonomy. In other words, for those who reside within its makeshift borders, the tent city functions, through the place-making tactics of its residents, as a space where the homeless can be "at home."

Tent cities were the brainchild of SHARE (Seattle Housing and Resource Effort), a small group of homeless and formerly homeless individuals. They emerged as a product of the dramatic political, economic, and social changes that were reshaping America's urban spaces in the 1980s and early 1990s. As noted in previous chapters, Seattle boomed from a small industrial port city to a "new economy," global city during this era. That prosperity, however, was largely borne on the backs of Seattle's poor and homeless, many of whom were the detritus of the region's waning industrial, agricultural, and forestry sectors. As the city rode the wave of technology-driven wealth, the very discourses of safety, cleanliness, and livability that attracted young, educated, urban professionals to the city's gentrifying core served simultaneously to justify the removal of the "new" homeless from Seattle's streets.

Responding to these circumstances in 1990, SHARE erected the first tent city south of downtown. It was near the site of what had been a depression-era Hooverville and was at the time home to the Olympics-style Goodwill Games. This original tent city served as a pragmatic response for a bourgeoning homeless population pushed onto the streets by federal welfare rollbacks, decreased affordable housing stock, skyrocketing rents, and punitive civility laws aimed at removing the homeless from public view. It also functioned as a symbolic gesture to render visible the plight of the homeless in Seattle to an increasingly affluent populace and an increasingly elitist-oriented city government.

After a series of "sanitation sweeps" by the city, SHARE founded Tent City 2 in 1998. Homeless encampments, along with the residents' belongings (and occasionally the homeless people themselves), were bulldozed from a wooded area south of downtown. After another unceremonious bulldozing by the city, Tent City 2 moved to private land donated by a local community center. After some legal wrangling over permits (which SHARE won), the city issued a consent decree allowing the Tent City to exist in perpetuity provided it was located only on private land and moved every ninety days. Tent City 3 was born. For the past eight years, Tent City 3 has continued its nomadic existence, moving every sixty to ninety days, mostly among Seattle's gentrified neighborhoods and inner-ring suburbs (see Appendix 3). Today, SHARE, joined by its offshoot WHEEL (Women's Housing Equality and Enhancement League), serves as the umbrella organization for thirteen indoor shelters and two tent cities.[5]

Tent cities are motivated by an overarching desire for an autonomous space in which residents can express and act upon their own needs and desires. The organization of a tent city foregrounds residents' voices in their collective struggle against the difficulties of homeless life, rather than having constantly to negotiate and struggle against common homeless stereotypes. Residents manage and maintain Tent City 3 entirely by themselves. They govern their space through weekly meetings, during which residents make and modify rules, plan community activities and outreach efforts, air grievances, and assign and volunteer for duties. Every activity, policy, and personnel change is subject to discussion and vote. Residents elect an "executive committee," as well as other officials, to serve in various coordinator and organizational positions. As camp founder Scott Morrow summarizes,

We wanted a way for homeless folks to come together as equals to stay sheltered and safe—in a way, it really boils down to self interest. Sometimes homeless people's best interests diverge from service providers. . . . When homeless people are the voters and the decision makers, it changes the dynamic in an important way.

The tent city's autonomy and self-governance stand in sharp contrast to the powerlessness of the street or shelter, as resident Nick describes:

Yeah, [in the shelters] they treat you like you're a little kid and after a while you become a little kid—it's really easy to fall into that "bum" roll if that's how everyone treats you. Here, you're responsible for something. If you don't like it, you can work to change it, if you don't do what you say you're gonna do, everyone suffers—yeah, it's a big deal!

Along with the benefits, of course, is simply the opportunity to be "at home." The importance of Tent City 3 as a home-like space was made clear to me the morning after my first night as a "camper." Having set up my tent using only the manufacturer-supplied rain-fly, I awoke the next morning cold and wet. From my soggy cocoon, I could hear a few muffled voices above the soft hiss of tires on the rain-soaked freeway below and the distant rumble of jets above. I groggily emerged from my tent and shuffled toward the smell of brewing coffee. When I returned, I noticed a few people standing around my tent with what looked like construction supplies. Apparently, my mistaken assumption that the stock rain cover would keep me dry was a common "newbie" mistake. The assembled crew assisted me in making a sturdy (and truly waterproof) rain cover out of tarps, 2 x 4s, and rope. After completing what felt to me like an old-fashioned barn raising, the group and (a very thankful) I sat down at a nearby table, where, amidst stories of other newbie gaffs they had made or witnessed, they explained to me how to use milk crates, cardboard, and other found objects to make the inside of my tent more comfortable and organized.

6.26 Example of common place-making tactics.

"Yeah" one woman explained, "this place ain't much, but we try to make it as much like home as possible."

A half an hour or so later, I sat pondering this statement as life in Tent City 3 went humming along about me. Most people were awake now and the communal area of the camp was bustling with activity. Some people were just leaving the camp for work, others were reading newspapers over coffee that had been bought and donated by other camp members. The drizzle had taken a brief hiatus and a few folks were munching day-old baked goods donated by a local Starbucks, while others completed their chores. One or two guys sharing a smoke at one of the tables laughed loudly. As I gazed out over this decidedly homey scene, it occurred to me that having a "place" in this world means far more than a roof over one's head. It means being able to rearrange the furniture, both literally and metaphorically— to *produce and create* space. One must not only have a space where one can have a modicum of control over one's possessions, thoughts, and actions, but also one must have the stability and security to know that when one leaves and returns everything will still be there—and the reassurance that there will be others there to welcome you back.

Motivated by the current economic recession and a recent spate of sanitation sweeps, a group of homeless and formerly homeless individuals have sought to create a more permanent alternative to the roving tent city. They call their camp "Nickelsville" after former Seattle Mayor Greg Nickels. They have met with constant resistance from city officials and have had to move from place to place in a quest for a permanent location. Claiming that the "Nickelodians" (their self-designation) are "merely protestors," the city has used zoning codes and trespassing ordinances to evict the Nickelodians and dismantle and confiscate their belongings. Protest or not, the fact remains that there are far more people who lack shelter than there are beds available and all signs point to this trend continuing in 2010.[6] Although the city and county have a plan "to end homelessness" by 2015, until then, residents of Seattle's tent encampments, if allowed, will continue to make living in a tent in a vacant lot "as much like home as possible."[7]

TURNING BACK THE CLOCK:
THE RESEGREGATION OF SEATTLE PUBLIC SCHOOLS

Tricia Ruiz and Mark Ellis

Seattle's racial history is replete with stories of both accommodation and repression. This is especially true around schools. There was never de jure segregation in Seattle; for many years, blacks attended majority white schools, and there is evidence of black parent involvement in PTAs (figs 6.27 and 6.28). However, by 1978 there was enough racial segregation in the district that the "The Seattle Plan" was implemented, which bused students across the city to achieve desegregation. It survived referendum repeal through a Supreme Court ruling in 1982. Pressure from some parents continued, and in 1988 Seattle ended mandatory busing and replaced it with a controlled-choice plan. This open-enrollment plan included a racial tiebreaker, which was used to determine school assignments, alongside factors such as siblings already enrolled and distance between home and school. Some white parents sued over the plan, and in 2007, the Supreme Court ruled that Seattle schools' limited use of race in deciding school assignments was unconstitutional.[8] The consideration of race in

6.27 Children at Seattle's Bailey
Gatzert Elementary School, 1943.

6.28 Bailey Gatzert School PTA
officers, 1946.

deciding admissions was the last remaining component
of Seattle's school desegregation policy. The most pressing question after the 2007 Supreme Court decision was
whether Seattle schools have resegregated.[9] We attempt to
answer that question here.

The Beginnings of Resegregation

To study the racial distribution across Seattle's public
schools we measured the percent minority by school clusters within the city (fig. 6.29).[10] From these graphs (figs.

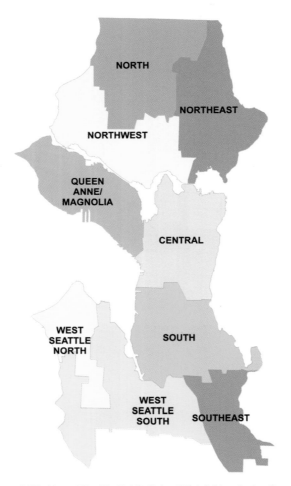

6.29 Map of Seattle Public School District by cluster. Source: Seattle Public Schools.

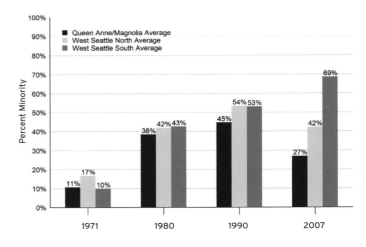

6.30 Percent minority in public schools for Queen Anne and West Seattle, based on Seattle public school enrollment data.

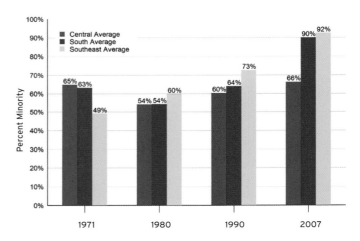

6.31 Percent minority in public schools for Central and South Seattle, based on Seattle public school enrollment data.

6.30–31), one can see that areas with increasing proportions of minority residents over time also had increasing proportions of minority students in neighborhood schools. As Seattle's North, Central, South, and Southwest neighborhoods became less white (see figs. 6.32–33), their schools also became less white. What is particularly telling is that there is a distinct decrease in the schools' average percent minority for Queen Anne, West Seattle North, Northeast, and Northwest neighborhoods (mostly white and prosperous areas) when comparing 1990 with 2007 (see, in particular, figs. 6.30 and 6.31). This time frame suggests that perhaps the end of busing and the elimination

of the racial tiebreaker in school assignments may have led to the decrease in minority population at schools in these four neighborhoods; while schools in the adjacent neighborhoods of the North, West Seattle South, South, and Southeast saw their nonwhite proportion grow.

Local Demographic Trends

Suburbanization, regional economic booms and busts, immigration, and changes in family size and household dynamics also have worked to change the region's popula-

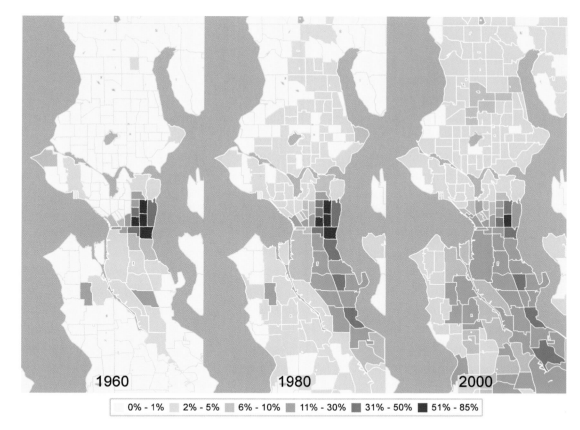

0% - 1%   2% - 5%   6% - 10%   11% - 30%   31% - 50%   51% - 85%

1960   1980   2000

6.32 Percent black population in Seattle, 1960, 1980, and 2000. Based on maps originally created by the Seattle Civil Rights and Labor History Project.

6.33 Percent Asian American and American Indian population in Seattle, 1960, 1980, and 2000. Based on maps originally created by the Seattle Civil Rights and Labor History Project.

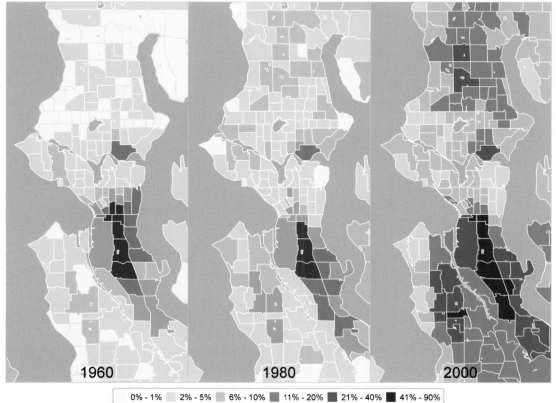

1960   1980   2000

0% - 1%   2% - 5%   6% - 10%   11% - 20%   21% - 40%   41% - 90%

tion structures, and thereby affect resegregation. From 1950 to 2000, as the Puget Sound region's population increased, King County's share of the regional population actually decreased.[11] And even within King County the suburbs grew at a much faster rate than the core city they surrounded. The relocation of families to the suburbs and the post–baby boom decline in fertility likely account for the majority of the decrease in Seattle's school-aged population.

Other data, however, indicate that white flight may also be partly responsible. Figures 6.34 and 6.35 show that the decline in overall enrollment within the city dovetailed, with a significant increase in minority students. By 1984, and since then, Seattle has been a majority minority district.

Part of this increased diversity is due to the rise in immigration to Seattle. By 2000 almost one in five Seattle residents was of foreign-born descent, which was 40 percent higher than it was in 1990. Over half of the foreign-born in Seattle are from Asia, although between 1990 and 2000, there was increased immigration from Africa and the Americas, too. In 2000 almost half of the foreign-born population in Seattle had arrived in the previous ten years.[12]

Differential birthrates likely also account for some of the growth in nonwhite student enrollment. Based on Census 2000 data, we know that the average family size for households headed by whites in Seattle is only 2.66. For households headed by blacks, the average family size is 3.18; for Hispanic households, it is 3.49; for Asian households, 3.47; and for Native American households, 3.02. These differentials, in conjunction with immigration, help drive change in the racial proportions of the local school-aged demographic.

Seattle's public school enrollment gap between whites and minorities stands in stark contrast to its overall demographic structure. Although the racial composition of Seattle is changing, it is still a much whiter and relatively wealthier city than one would expect by looking at the public school population. Indeed, a local headline story titled "A Tale of Two Seattles" illustrates how Seattle's economic boom between 1996 and 2006 paradoxically paralleled the increase of minorities and poor students in local schools.[13] This article illustrated that the average Seattleite was more likely to be white and middle-class compared to the average Seattle public school student, who was more likely to be a racial/ethnic minority and more likely to qualify for free or reduced lunch.

So where are the missing white middle-class students who come from the families that have prospered in Seattle

6.34 Seattle Public Schools majority and minority enrollment, 1968-2002, based on Seattle Public Schools enrollment data.

6.35 Seattle Public Schools enrollment by race/ethnicity, 1976-2002, based on Seattle Public Schools enrollment data.

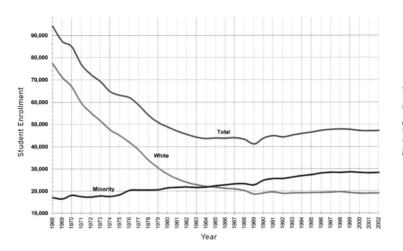

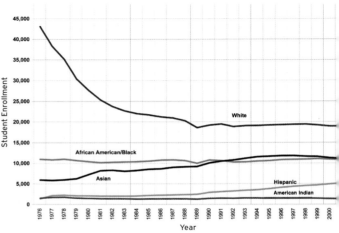

over the last twenty or so years? Private school enrollment is the key to answering this question. To understand differential rates of public school attendance in Seattle by racial group, one must consider geographic patterns of private school enrollment. The availability, proximity, and quality of private schools can greatly determine whether local public schools have to "compete" for the local market share of school-aged children, in addition to parents' level of income, wealth, or even religious preferences. Seattle has had a much higher private school attendance rate than the rest of King County, as well as other nearby counties. During the last ten years, approximately 25 percent of K-12 students in Seattle attended private school, compared to 5 percent in Snohomish County, and 10 percent or less in Pierce County and in King County not including Seattle.

There is also an acute racial gap in private school attendance in the city compared to surrounding areas. In 1999 more than 70 percent of Seattle's private school students were white, and although this number decreased slightly to just over 60 percent in 2007, this picture stands in marked contrast to Seattle's public schools, where whites accounted for only about 40 percent of students. In other words, Seattle not only has a higher than normal rate of private school attendance, but also those who do attend private schools are much more likely to be white. Thus, public schools in Seattle have a much smaller white share of students compared to other public school districts in the area.

Geographies of Public School Attendance

Public school choices in the city are largely determined by where a student lives, unless parents can afford to send their children to private school or provide home schooling. Where in the city are these options most likely to be taken? And who takes them in those areas? Does the racial configuration of neighborhoods correlate with decisions?

For figures 6.36–39, we used Census 2000 data and 1999–2000 enrollment data from Seattle Public Schools. We calculated and mapped local rates of public school

attendance at the block-group level (a Census-defined boundary). The rate is simply defined as the number of students enrolled in public schools divided by the number of school-aged children who live in a block group. First, in figure 6.36, we map which areas of Seattle have higher rates of public school attendance. We see that there are much higher proportions of school-aged children attending public schools in the south end of the city—the area where nonwhite populations congregate—compared to the neighborhoods to the north and west. We also map the racial differences in these public school attendance rates in figures 6.37–38.

For nonwhite students, the overall distribution is somewhat even. Whether they live in white or nonwhite majority parts of the city, nonwhite students attend public

6.36 Proportion of school-age population attending public schools, based on Census 2000 data and 1999-2000 enrollment data from Seattle Public Schools.

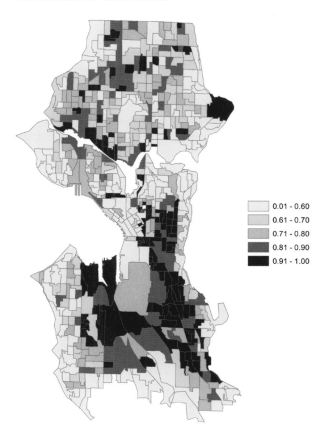

	0.01 - 0.60
	0.61 - 0.70
	0.71 - 0.80
	0.81 - 0.90
	0.91 - 1.00

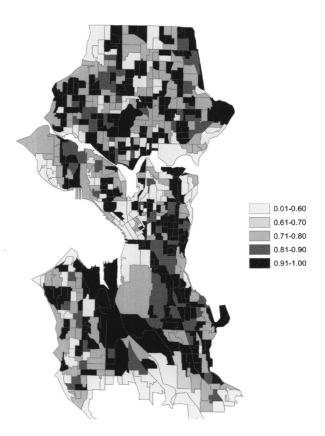

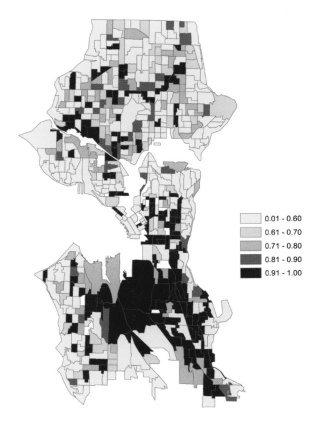

6.37 Proportion of nonwhite school-age population attending public schools, 2000, based on Census 2000 data and 1999–2000 enrollment data from Seattle Public Schools.

6.38 Proportion of white school-age population attending public schools, 2000, based on Census 2000 data and 1999–2000 enrollment data from Seattle Public Schools.

6.39 Difference between white and nonwhite public school attendance rates, 2000, based on Census 2000 data and 1999–2000 enrollment data from Seattle Public Schools.

0.01-0.60
0.61-0.70
0.71-0.80
0.81-0.90
0.91-1.00

0.01 - 0.60
0.61 - 0.70
0.71 - 0.80
0.81 - 0.90
0.91 - 1.00

whites lower rate than nonwhites

whites same rate as nonwhites

whites higher rate than nonwhites

school at the same rate. But for white students, the geographic unevenness is stark. White students in the north, west and outlying areas of the city attend public schools at a much lower rate compared to white students in the south end of Seattle. White public school attendance is thus highest in parts of Seattle where whites are in the minority. Where whites are in the majority in the north—the area of the city in which are found many of the best public schools—white public school attendance rates are lower.

In figure 6.39, we calculate the difference between the two rates for white and nonwhite students. Again, we are able to see that in general, white students attend public schools at much lower rates than nonwhite students except for the few pockets in the north end and in the general middle corridor of the south end.

Back to the Future

Partly in response to the most recent Supreme Court decision and after decades of busing and controlled-choice school enrollment, in 2010 Seattle public schools began a new neighborhood-based assignment plan. This plan eliminated almost all aspects of the previous choice-based system and returned the city's schools to an assignment scheme much like the one that existed in the pre–*Seattle Plan* era. This "back to the future" change occurred in the context of increasing public school enrollment at most grade levels. From 2005 to 2009, overall public school enrollment applications increased by 10 percent. Applications for pre-K were up 26 percent, K–5 increased 15 percent and grades 6–8 rose by 14 percent. Grades 9–12 slightly decreased by 1 percent.[14] What do these enrollment trends and new neighborhood attendance plans portend for segregation and diversity in Seattle schools?

If the increase in enrollments primarily occurs in the north end where there are higher proportions of school-aged children who are white, then public schools in the North district will have an increase in white enrollment. And the inverse is also possible; if enrollments increase in the Central and South districts where the majority is

nonwhite, then neighborhood schools there will see even higher rates of nonwhite attendance. Without more detailed current data, we can only speculate about the main driving factor behind increased school enrollments and from where in the city these students will most likely come.

Until some time has passed after the new neighborhood-based enrollment plan has been in effect, and until we are able to document more clearly the geography of financial hardships experienced by families in Seattle, we will not have the tools or data to answer these questions properly. However, what we do know is that Seattle will continue to struggle with segregation and lack of diversity in its schools as long as attendance is tied to its segregated neighborhood spaces. The future looks very much like the past for Seattle's public schools.

FOR WHOM DID SEATTLE SCHOOL DESEGREGATION WORK?

Catherine Veninga

When academics and policy makers ask whether or not school desegregation "worked," they usually try to determine if it addressed the key assertions in the pivotal *Brown vs. Board of Education* decision: that a racially segregated educational system (1) provided black students a substandard education because it denied them access to the same academic resources as white students, and (2) generated a sense of inferiority among black students. A third question that builds on the first two regards parity of opportunity: Did desegregation improve the "life chances" of blacks? Because busing policies were ostensibly implemented in order to guarantee the constitutional rights of black students by providing them access to educational opportunities that whites already enjoyed, desegregation has been considered, by and large, a policy *for* blacks.

My research on busing in Seattle, however, suggests that in important ways, school desegregation was also a policy *for* white students.[15] My study was based on interviews with individuals who had participated in Seattle's desegregation program between 1978 and 1990; instead of asking questions about academic achievement, I was more interested in what the experience was like. I wanted to learn—from the perspective of participants—what it meant to be bused. I found that one of the most formative and valued aspects of the busing experience was exposure to difference—not just class and racial difference but difference in types of music, dance, food, landscape, values, and styles. For many students, direct and daily exposure to such diversity taught them to challenge popular stereotypes, "broaden their world view," and engage difference rather than avoid it. Respondents often talked about the experience of exposure as a "social education" and described it as something that "soaked in" over time; it was not part of a lesson plan nor did it show up on a final exam.

That's what integration was about. It wasn't about us sitting in a room and talking about integration every day—that wouldn't have done anything, it would have just fired us all up and everybody would have had their guards up. So instead, what changed something? It changed because we were together. We didn't talk about it, we were together.

While the impact of exposure was significant for students from all racial backgrounds, it may have been most profound among white students. Many white people fail to realize that as a racial group, they are more spatially and socially isolated than members of other racial groups. This was true in Seattle during the 1970s and '80s when whites made up over 80 percent of the population and lived in neighborhoods where they were unlikely to encounter people of color. Because the nonwhite population was much smaller, it was less true for blacks, and not true for Asians, who were very likely to encounter members of other racial groups in their neighborhoods. Therefore, the experience that many white students had of attending a racially diverse school was much more of a "shock" (as many respondents described it) than it was for students of color. Moreover, as one respondent said, "minority kids grow up

seeing a white world"—we live in a country where our basic social norms and values are defined in terms of whiteness.

People of color have to deal with white people in this society. White people don't have to deal with people of color, and so for them to get a chance to makes them lucky. I think that people of color sort of get to see what white people do, I don't know if it works so much the other way.

Busing for school desegregation was an important social policy not only because it redistributed academic resources but also, and perhaps more importantly, because it redistributed people. By focusing exclusively on how desegregation impacted black achievement, scholars and policy makers miss the opportunity to explore how it may have altered whites' attitudes regarding racial difference. This is especially significant in a city like Seattle, where whites have historically sought to isolate themselves from people of color via various social institutions, and perhaps most profoundly, through the housing market.

SPATIAL STORIES:
BELLTOWN, DENNY HILL, AND PIKE PLACE MARKET

Kim England

Asking *for whom* cities are created is central to urban social geography. Urban neighborhoods are produced through a multitude of processes, practices, and issues as they come together in particular ways in particular cities. Since I've lived in Seattle, I've looked to the immediate northwest of the central business and commercial district to find places to explore those processes (fig. 6.40).[16] For me, this area serves as one of those illustrations that the famous urbanist Jane Jacobs urged us to look for: "For illustrations, please look closely at real cities. While you are looking, you might as well also listen, linger, and think about what you see."[17] For the last ten years I've had students in my Geography 277: "Geographies of Cities" course go and look closely at (and listen, linger, and think about) Belltown. Such field-

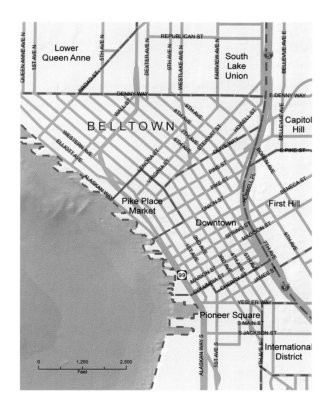

6.40 Belltown.

work is a long-standing practice in urban geography. This area has a colorful history—one of unfolding spatial stories that illuminate the relationships among places, people, and things. The stories of Pike Place Market, Belltown, and Denny Hill are about economic restructuring, politics, and urban planning, as well as about power, greed, and marginalization. The area is shaped by and has shaped Seattle's versions of "urban reclamation," "urban renewal," "urban revitalization," and "urban boosterism."

Pike Place Market is usually dubbed one of the top places to visit in Seattle (fig. 6.41). It has been in continuous operation since 1907, attracting millions of visitors a year. Its story includes chapters on the internment of the Japanese Americans in Washington State and the fight against urban renewal. From the first days of the Market until the Second World War, Japanese and Japanese American farmers were a significant presence among the suppliers and vendors of farm produce. In the early 1940s,

Japanese American farmers supplied or ran four-fifths of the Market stalls. Executive Orders 9066 and 9095 forced their removal from Seattle and internment in camps during the Second World War. Their removal was a contributing factor to the post-war decline of the Market, along with the development of neighborhood supermarkets, the suburbanization of the population, and the removal of truck farms from the city.

In the early 1960s, the City's "Pike Plaza Redevelopment Project" called for demolishing the existing Market and replacing it with high-rise apartments, hotels, and a seven-level parking garage for 6,000 cars (fig. 6.42). City Council member Wing Luke quietly urged city attorney Robert Ashley, architect Victor Steinbrueck, and various local groups to organize a public effort to protest the bulldozing of the Market. An urban social movement was born. An April 1969 poster declared "Citizens of Seattle and King County, in the name of common decency and the tradition and heritage of our region, the Friends of the Market summon you to our cause of keeping this low-cost market residing in the Pike Place Market area for all people for now and for the future. The merchants of greed are using urban renewal to murder the market." Later that year, the City Council voted to go ahead with the urban renewal plans. In 1971 Friends of the Market launched Initiative 1 to save the Market; 25,000 signatures were collected, securing a place on the November ballot. The citizen initiative saved the core of the Market. The City purchased the Corner Market Building in 1974 and rehabilitation began, creating the nine-acre Pike Place Market Historical District that ensured the survival of the Market (fig. 6.43).[18] These two moments in the Market's history are memorialized by the 1999 installation of seven porcelain, enameled panels called "Song of the Earth," by artist Aki Sogabe, near the entrance to the Market; and after Steinbrueck's death in 1985, the small park at the north end of the Market was renamed Victor Steinbrueck Park.

Farther to the north of Steinbrueck Park is Belltown. In 1997, *Utne Reader* ran a story called "The 15 Hippest Places to Live: The Coolest Neighborhoods in America and

6.41 Pike Place Market.

6.42 Pike Place Market urban renewal study, 1964.

6.43 Corner Market during rehabilitation, 1975.

Canada." Number 7 on the list was Seattle's Belltown. The piece begins by declaring that "by its very nature, hip is something ephemeral and ultimately indefinable. Yet you know it when you see it—by the way a place looks and feels."[19] Belltown makes it onto the list as "the incubator of artiness. Galleries, restaurants, and bars . . . share the streets with human-scale apartment buildings." Generations of boosters and promoters of this neighborhood, one of the oldest in Seattle, would have given a huge collective cheer that at last their vision had borne fruit. Even today Belltown continues to be trumpeted as a hip place. The Web site "All about Belltown" still claims that "Belltown is the epicenter of Seattle's restaurant scene, with hot nightclubs, cool shops, great places to live, work or play, the new Olympic Sculpture Park . . . Belltown's got it all!"[20]

In more ways than this Web site infers, Belltown *does* have it all. Amid these newer landscapes of wealth and

consumption are continued "landscapes of despair."[21] Belltown is a bewildering mix of wealth and poverty with half-million dollar condominiums (despite the current financial crisis) alongside the homeless. Census data analysis for the city of Seattle shows that the two census tracts for Belltown have the highest household and individual income inequality in the city. And that inequality is clear on the ground. On Western Avenue, for example, on the block between Vine and Cedar streets is the Vine Building, a ten-year-old condominium building in a renovated warehouse, where a 1 bedroom/1.5 bath condo is on sale for $350,000. A few steps down along Western, toward the downtown core is the Millionair Club Charity, founded in 1921 with the goal of providing jobs and direct assistance to people in need. And farther along Western, by the overpass, there used to be a Latino day-worker center and brightly painted wall murals, including one that proclaimed "Work Is Progress." The bilingual signs warning "No Trespassing" and "No Loitering" are still there (fig. 6.44).

The roots of what are now Belltown and Denny Triangle go back beyond the arrival of Seattle's first Europeans settlers. The pre-white-contact area was dotted with Duwamish encampments and was a valued site for hunting deer and elk. Until the early twentieth century, the waterfront remained an important camping place for coastal Native peoples traveling to and from Alaska and British Columbia and along the Washington coast.[22] William and Sarah Bell and their children were part of the original Denny party that landed on Alki beach in 1851. Bell's original 1853 claim was a densely forested area that ran

6.44 No Trespassing sign under the Viaduct at Bell Street.

6.45 Austin A. Bell Building.

north to south from present-day Denny Way to Union Street, and east from the bluffs overlooking Elliott Bay (already studded with longhouses and camping grounds) over a hill to a point beyond the location of Interstate 5. In 1870 Bell focused his attention on the waterfront portion of his claim, in what is now considered Belltown, on a stretch on First Ave east of Battery Street. His son, Austin, inherited the Belltown properties and hoped that new construction, such as the 1889 red-brick Austin A. Bell Building (fig. 6.45), still standing between Bell and Battery, would help establish Belltown as a commercial center. That was not to be. At one point the Bell Building was a low-rent apartment hotel, and after that it sat vacant for decades.

The processes and practices that led Belltown to become one of the "hippest places" ensured that the Bell Building was renovated into luxury condominiums with street-level commercial use. In addition, the building was designated a historic landmark, even though the renovation involved the often maligned practice of facadism

(gutting the interior and retaining only the front brick facade). The legacy of the Bell family remains fixed on the landscape, not only through this building but also because nearby Bell, Virginia, Olive, and Stewart streets are all named for members of the Bell family. On the other hand, there is no longer any evidence of the Duwamish encampments that were once scattered over the area.

Reflecting on the physical landscape that makes up this part of Seattle, historian Coll Trush remarks that "Seattle is a bad place to build a city. Steep hills of crumbling sand atop slippery clay . . . ice-age kettle lakes and bogs, and plunging ravines and creeks are all sandwiched between Puget Sound and vast, deep Lake Washington."[23] One of the offending hills, named for one of the city's founders, was Denny Hill, which covered approximately sixty city blocks (fig. 6.46). On its southern peak, the imposing Denny Hotel stood from 1890 to 1906 (by then under new ownership and called Washington Hotel). The hill fell steeply from the highest peak at Second Avenue and Leonora Street toward the waterfront of Belltown. The leveling of Denny Hill began informally in the 1890s, but the city got involved in the early twentieth century. The scale of the project got ramped up under the management of the formidable city engineer, Reginald Heber Thompson, and preceded in two stages in 1902–10 and 1929–30.

In 1916 local historian Clarence Bagley described Seattle's business area (in which he included Belltown and the Denny Triangle) as "virtually one vast reclamation site," which captured both the chaos and the excitement of the time. Many embraced the regrading of Denny Hill (and other hills in Seattle) as an exemplar of Progressivism. A collage of photos of the regrading appears in Bagley's book under the title "Denny Hill Regrading: Moving 750,000 cubic feet of earth to build a city."[24] City leaders and private landowners were told that leveling the hill would increase land values. And at least at first it did. For a time there were even visions that Seattle's commercial district and civic center would move north to brand new Beaux-Arts buildings based on the "City Beautiful" model (a grand civic-plaza design by Virgil Bogue, a former colleague of the

6.46 Belltown, 1882, looking north from the top of Denny Hill; the street is Second Avenue.

Olmsted brothers who had been hired by the city to create the model). In 1912, Seattle voters rejected that plan.

The 1902–10 regrade focused on the waterfront side of Denny Hill from about Fifth Avenue. The second wave of construction focused on what remained of the hill east of Fifth. Engineers sluiced some of the hill into Elliott Bay; some was shoveled up and moved to create Harbor Island. As a "planning solution" the flattening of Denny Hill was heralded as a momentous civil engineering feat. Homeowners who did not move had the hill sluiced around their property, leaving the homes sitting on top of a pile of dirt, in some case requiring a stepladder to reach the front door. The regrading process did meet with resistance, and from time to time was halted. More than forty "regrade cases" reached the state Supreme Court. Even some of the original supporters of the regrade later revisited their position. In his memoirs, Thompson defiantly remarked, "Some people seemed to think that because there were hills in Seattle originally, some of them ought to be left there."[25]

In 1917 the Belltown/Denny Regrade area included some of the most prominent buildings to be found in Seattle, along with some less prominent ones on the level-

but-empty thirty or so city blocks. A map from the era reads, "The growth of Seattle made necessary the removal of the hill and the establishment of an even grade. This has recently been completed with the streets all graded and paved. The remainder of this hill from Fifth Avenue to Thomas Street has yet to be removed." Between 1928 and 1931, the rest of the hill was also regraded, when about thirty-eight city blocks were leveled and the dirt was carried to Elliott Bay on large conveyor belts (fig. 6.47). In the end, about sixty-two city blocks were affected.

The Regrade District and Belltown never became part of Seattle's commercial center as city officials had hoped. Instead, this area became home to light industry and other land uses that did not need a prestigious address in the downtown core. By the mid-1940s, the entire Denny Triangle area, including Belltown, and the Pike Place Market, had been designated "transitional" by Calvin Schmid, the University of Washington sociologist who developed typologies of "natural areas" in Seattle along the lines of

Chicago's urban ecologists.[26] Through the post-war era, the district became home to small offices and warehouses. Belltown, and to some extent the Market, later became popular with artists because of cheap rents and extensive living spaces that could be used as art studios. Other spaces became clubs that helped promote Seattle's grunge rock scene. The Belltown that the *Utne* article described was the result of a period when expensive condominiums were replacing inexpensive studios. Increasingly spaces were

just as likely to be occupied by designer boutiques and fancy restaurants as by light industry, artists, thrift stores, or alternative-music nightclubs.

GENTRIFICATION AND THE "STAYERS" OF COLUMBIA CITY

Gary Simonson

Along Columbia City's five-block commercial core in southeast Seattle, the car-oriented Rainier Avenue strip suddenly adopts a pedestrian scale: smaller blocks, wid-

6.47 Before and after of the Denny Regrade in progress from 1928 to 1931.

ened sidewalks, and many crosswalks (fig. 6.48). The historical character of the late nineteenth- and early twentieth-century buildings (virtually nonexistent in the rest of Rainier Valley) has been preserved. Boutique shops, upscale "cosmopolitan" restaurants, pubs, ethnic eateries, a wine bar, an art gallery, and a cabaret theater line Rainier Avenue and its side streets. Condominum developments are scattered throughout the immediate area. Fifteen years ago, Columbia City looked nothing like this, but gentrification has brought about remarkable changes in the neighborhood.

Within popular culture and mainstream media, the term *gentrification* has become increasingly associated with notions of urban revitalization and urban renaissance. Yet, it is important to remember that the term was originally coined to describe a *class-based* transformation. It is fundamentally the reclamation of the inner-city by the middle class (away from the working class). Hence, in order to gain a better understanding of Columbia City's changing social geographies, I sought to document the class-based transformation occurring there.

Most academic research on the effects of gentrification has examined residential "displacement," when working-class residents are forced to move out of their homes as a result of the process. However, I examined the impact of gentrification on working-class residents who *remain* in Columbia City (or "stayers," as I call them). Have those who remained benefited, as gentrification rhetoric often implies, from what the process has brought to the neighborhood? Are the supposed "improvements" and "amenities" appreciated?

First settled in the late nineteenth century and annexed by Seattle in 1907, Columbia City was a prosperous, middle-class, racially homogenous, white neighborhood until the late 1960s.[27] By the early 1970s, however, the end of legal racial covenants and the Great Boeing Bust (when Boeing laid off over two-thirds of its 100,000 employees) brought about dramatic changes to Columbia City and the rest of Rainier Valley.[28] An influx of African American—and to a lesser extent, Asian American—residents

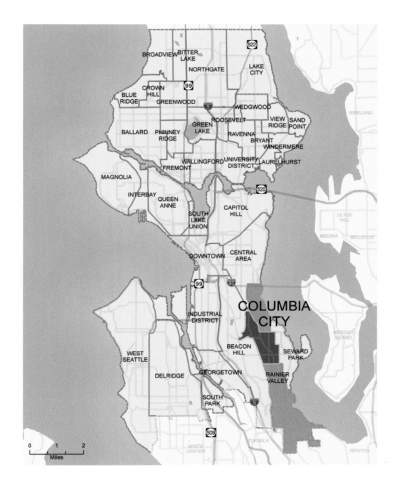

6.48 Columbia City in Seattle.

into the neighborhood, accompanied by "white flight" and increased unemployment, marked the beginning of Columbia City's transition into one of the city's poorest neighborhoods.[29] Over the next two decades, investment fell as the neighborhood's racial profile changed.[30]

By the late 1980s, Columbia City (along with the rest of the Rainier Valley) had become known throughout Seattle as an area that was riddled with violence, prostitution, and drugs. Poverty rates soared, and most residents did not have significant disposable income.[31] Subsequently, many businesses along the commercial strip were forced to close, creating a landscape littered with empty storefronts and abandoned buildings. However, some local establishments—such as churches, hair salons, and restaurants—did remain in Columbia City during that time, most of them

6.49 The "revitalized" Columbia City commercial district. Tutta Bella Pizzeria, Columbia City Theatre, and Columbia Alehouse near the corner of Rainier and Hudson.

catering to the predominantly African American community. Indeed, by 1990 Rainier Valley was the area with the highest concentration of African Americans in Seattle.[32] Despite urban decay and disinvestment, Columbia City had become an important hub for Seattle's African American community.

During the 1990s, a state-facilitated gentrification process took shape in Columbia City. The Growth Management Act led to the creation of Seattle's "Neighborhood Planning Program."[33] This program allowed residents to develop their own neighborhood plan (within the framework of the city's vision), while the city provided them with technical and financial support. In Columbia City this program had rather neoliberal effects. A handful of middle-class residents and potential business owners devised a comprehensive plan that included a variety of "growth-oriented" goals and policy proposals.[34] The plan

was implemented by the city in 1999 and has since facilitated the redevelopment of Columbia City's commercial and residential areas.

During the early-to-mid 2000s, different types of businesses appeared along Rainier Avenue, including a variety of new restaurants, pubs, boutiques, and entertainment venues (fig. 6.49). Columbia City's residential landscape changed dramatically during this time, with many apartment buildings being transformed into condos and housing prices rising significantly. The neighborhood shifted demographically, too, with most accounts pointing to the African American population declining considerably, and, conversely, to the white, middle-class population increasing.[35] Still, despite this influx of middle-class residents, there remained a significant working-class, minority population in Columbia City through the late 2000s.

I interviewed twenty-one residents and business owners, approximately half of whom were categorized as "stayers."[36] The remaining respondents were categorized as "gentrifiers," "business owners," or "other." Incidentally, all stayer-respondents (aside from one Asian American) were African American, while all gentrifiers were white. The interview questions were straightforward and focused on "neighborhood change" in Columbia City.

Overall gentrification has had mainly negative impacts on the stayer community there. Stayers argued that the redevelopment of the commercial strip has been geared toward middle-class gentrifiers, while older, stayer-oriented establishments have increasingly disappeared. In addition, stayers have virtually no say in neighborhood decision-making processes.

Many stayers also feel isolated by class-based shifts in social, economic, and cultural norms and have lost their sense of belonging in the neighborhood. Stayers in Columbia City have not only been "blanked out" by gentrification, their particular needs and values have also been positioned as a nuisance to the development of the neighborhood. Yet, despite all of this, their physical presence provides important *rhetorical* social value for "progressive" middle-class residents, who are happy to proclaim that they live in a "diverse" and "authentic" neighborhood.

Less Affordable, Less Accessible

The newer establishments that have recently sprung up in Columbia City are either not affordable or not culturally accessible for many stayers. Most stayers expressed concern that the new shops and restaurants were too expensive and could only (at best) be enjoyed as a rare treat. Some stayers emphasized that these establishments were clearly catering to newer middle-class residents and not "regular people" (i.e., the stayer community). As one stayer put it, "Who's going to pay $10 for a hamburger? The white folks paying that, not us!"

Respondents also discussed how many of the newer establishments are not culturally accessible for stayers. Stayers do not appear to share the same rhetorical and consumer-driven preferences for the "cosmopolitan" and the "historical" that have defined the new middle class. Thus, establishments such as the Columbia City Art Gallery, the Verve Wine Bar, and Tutta Bella Pizzeria appeal primarily to middle-class residents. One stayer's description of Tutta Bella (a popular "authentic" Italian pizzeria with a rare foreign certification) is illustrative of this point:

The food has changed. The food seems more like not real down home food like it used to be. . . . I'd say more like hamster food style. You know, the pizza, for example, it's supposed to be really good, and it's really not that good. Not much meat or cheese. Put some meat and some cheese on your pizza, you know? . . . Their pizza—I'm more of a Pizza Hut guy, you know?

This is just one example that highlights the fact that stayers do not necessarily have the same cultural sensibilities as gentrifiers. In this case, this man does not want "authenticity," he wants a place that is hearty and familiar.

Ultimately, newer establishments brought about by gentrification have not benefited Columbia City's resi-

dents equally. Groups with very different income levels and cultural sensibilities share the neighborhood, but the recent development of the commercial strip has been geared almost entirely toward newer middle-class residents. Although proponents of gentrification often speak of increased amenities for all, the plethora of new boutique-type, cosmopolitan establishments have not necessarily provided such amenities to stayers.

Lost Establishments

Many of the establishments that catered to stayers and helped tie their community together have been displaced since the onset of gentrification. Despite the fact that "pregentrification" Columbia City was marred by urban decay and abandoned buildings, there were a range of establishments that served as an integral and important element of the residential community at that time. Most of these establishments are now gone. As one stayer put it: "While we used to have the mom and pop stores and we used to have everything in our community, we don't have that any longer, what a normal, functional community should have. If you're taking all that away from us, then what do we have?"

According to many respondents, the economic, cultural, and social pressures of gentrification triggered the disappearance of these establishments. Many simply could not afford to pay increasing rents or property taxes, and were often targeted by commercial developers who believed the space could be occupied by more profitable businesses. In addition, some of the older businesses were viewed by newer residents and business owners as being cultural and economic liabilities (due to "unsavory crowds" and "run-down" aesthetics) that were detrimental to the overall reputation of the new commercial strip. These places were ultimately not part of the vision certain middle-class gentrifiers (and developers) had as to what the neighborhood should be.

Stayers recounted numerous examples of how outside social pressures forced certain establishments out. One case involved a long-time community church that was forced to sell to developers as a result of increased property taxes and a dwindling congregation. The church had previously attempted to renovate their building (to attract new members), but had been blocked by the Columbia City Landmark Review Board (an arm of the city government), which claimed that the historical integrity of the building would be violated. Once developers bought the building, however, the Review Board allowed them to gut and transform the space into several commercial businesses.

Another example involved the Columbia Plaza, which had been one of the last affordable retail shops in the neighborhood. It was vilified by much of the newer business community as being a place that attracted violence and other "bad behavior," ultimately leading to its lease not being renewed. The space is now slated to become condos and retail space. Unfortunately, examples such as these demonstrate how stayers in Columbia City have lost both practical places to shop *and* important community-oriented hubs.

Pushed to the Fringes

Many stayers feel isolated and claim that gentrification has pushed them to the physical and the symbolic fringes of the neighborhood. Obvious economic discrepancies and cultural differences between stayers and gentrifiers, as well as racial tensions, cause certain stayers to feel as though they are now "out of place," despite the fact that they have lived in the neighborhood longer than most middle-class residents. As one stayer says, "When I'm walking down Rainier Avenue, they're walking around doing their own thing and stuff, and I don't know, it makes me feel out of place, like I don't belong here. And I've been here a lot longer. If they had any idea—a lot longer."

This quote, among others, emphasizes that many stayers feel uncomfortable walking through the heart of the neighborhood (the main strip of Rainier Avenue). As one long-time business owner points out, much of this also has to do with the widening gap in the social geographies of Columbia City:

Those folks [stayers] don't wander down this way so often anymore. They're just back in there [in the side streets], you know. We know where they're at. But they're just off the road and down into these little holes, apartment complexes that are back in there, and so, those folks, they're around but you don't see them up and down the neighborhood.

The "holes" he refers to are run-down, low-income apartment building clusters on streets surrounding Rainier Avenue. They may be in close physical proximity to the main commercial strip, but they are hidden from view and are becoming increasingly isolated (both socially and culturally) from the now-bourgeois core of the neighborhood, shifting the spatial scale of segregation in Columbia City to a micro-level.

Along with decreased visibility, stayers also appear to have very little influence on neighborhood decision-making processes. When middle-class residents began moving into Columbia City in larger numbers, informal social networks were formed through organized social gatherings (e.g., a monthly "Sunday potluck"), a neighborhood Web site where residents could negotiate needs and exchange information, and various other endeavors. Not surprisingly, these informal networks spilled over into more formal neighborhood organizations that have significant power and influence in neighborhood decision-making processes (the aforementioned Columbia City Landmark Review Board, the Columbia City Business Association, etc.). However, as a result of structural—although not necessarily intentional—exclusion from these networks and organizations, stayers have had minimal opportunities for meaningful civic engagement in their own neighborhood.

Ambivalence

Despite their overwhelming emphasis on the negative impacts, some stayers were surprisingly ambivalent when assessing how Columbia City has changed as a result of the gentrification process. Not surprisingly, there were some positive impacts mentioned (albeit, sparingly) by stayers. These included an increased sense of safety (by far the most valued positive impact), the improved aesthetics and reputation of the neighborhood, and renovations to the library and various parks.

Yet, unexpectedly, there were some stayers who—despite railing against the various ways in which gentrification has negatively affected their lives—believed that the neighborhood had *improved*, citing reasons that they acknowledge have primarily benefited others, not themselves. At times, the improvements cited by such respondents are, ironically, the direct causes of the negative impacts they criticize. For example, one stayer who earlier spoke of his discomfort walking down the commercial strip since gentrification, later applauded the fact that gentrification has made the strip more vibrant. He states, "It's great, it's bringing people out to, you know, get around the community. But it's making people like me, I don't know why, maybe it's just me, but there's a lot of me's, feel uncomfortable with it."

Another stayer, who previously stated that he believes the food in the neighborhood is no longer for "regular" people, later says, "But the food, I think the food has *helped* the neighborhood, as far as bringing more people into the neighborhood from other parts of Seattle."

How should we understand these ambivalences? Perhaps they are a result of conflation between the rhetoric that has accompanied the gentrification process and the reality that has emerged. These contradictory statements may stem from individuals' inability to reconcile what the media and the city say about gentrification with what they encounter on a daily basis. Pro-gentrification rhetoric has been successful in convincing some that it is an inevitable and "natural" process. This inevitability was expressed by respondents who, while acknowledging that gentrification has negatively affected certain people in the community (including, at times, themselves), also believe that it has been entirely unavoidable. For them, the process is viewed as part of a natural progression that is useless to resist. For example, one respondent laments that gentrification has

only benefited some people (in her view, "mostly rich white people") while marginalizing stayers, but she later insists that "change is good. You can't stop change."

This fatalism may be why there has been little organized resistance to gentrification in Columbia City. Considering the content of the interviews and the narratives presented by many respondents, one would imagine that there would be far more passionate denunciations of gentrification, particularly by stayers who have been most vulnerable to its negative impacts. However, aside from a couple of exceptions, this was not the case. Such ambivalence provides insight into why most stayer respondents in Columbia City appear to be more confused than angry about what has occurred in their neighborhood.

What Columbia City has shown us is that such a class-based transformation can have significant negative impacts *aside* from residential displacement, raising new questions and concerns about gentrification (or "urban revitalization") as a primary urban strategy. In Columbia City, even though gentrification has fostered a limited "mixed-class" community, it has been a restructuring of an inner-city neighborhood for middle-class consumption and habitation. Clearly, demographic shifts do not tell the whole story regarding impacts of the process on this community. Here, gentrification has also caused neighborhood-scale segregation, and stayers, as one respondent aptly put it, have been relegated to the role of "second-class citizens."

QUEERING GAY SPACE

Michael Brown, Sean Wang, and Larry Knopp

Ask anyone where the gay or queer area in Seattle is, and more than likely they'll say "Capitol Hill, of course!" This relatively dense, inner-city neighborhood in the center of the city is somewhat akin to districts like the Castro in San Francisco or Greenwich Village in New York. So it reflects a general trend in twentieth-century American cities, where sexuality and sexual identity became visible foundations of neighborhoods. Capitol Hill and other "gay ghettoes" reflected and reinforced the growing visibility of gay, lesbian, bisexual, transgender, and queer (GLBTQ) identities through economic, political, and cultural visibilities in the city. They also constituted formidable spatial resistances to homophobia and hetero-normativity in urban life, and in American society generally. These districts emerged in the latter third of the twentieth century and have been associated with gentrification and its complicated racial and class dimensions. They have increased visibility of GLBTQ people and culture in the city and they have become important spatial anchors for community and political change.

Social geography seeks to understand the relations and practices that constitute identities (like GLBTQ) and places (like urban neighborhoods). In the case of Capitol Hill, there

6.50 Map of gay venues in spring of 2010.

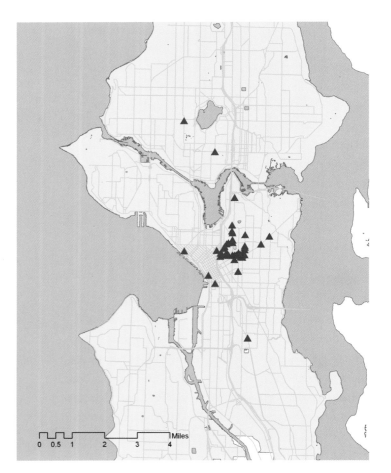

are many overlapping ways in which it and its residents are constituted as "queer." Capitol Hill is an *economically* queer district. The clustering of retail and commercial venues catering to GLBTQ folk (gay bars, nightclubs, restaurants, bookstores, religious groups, sex clubs, cruising areas, etc.) is clearly shown in figure 6.50. So we can say that commercial property's spatial concentration evinces a gay ghetto. But of course there's more to it than that.

Popular knowledge in Seattle holds that it is *residentially* populated by many who identify themselves as gay, lesbian, bisexual, transgendered, or queer. Gary Atkins's touchstone history, *Gay Seattle,* notes that through the 1970s and 1980s, the Hill's large number of affordable apartments and rooms in shared houses (due in no small part to white flight and fears of inner-city decay) reflected a rental market that drew young queer baby boomers into the area.[37] And indeed, the 2000 Census reports that Capitol Hill tracts have large percentages of same-sex couples (more on this later).

Capitol Hill is queer *politically*. It has been the site of several important political rallies, organizations, and nonprofits. Organizations such as the Northwest AIDS Foundation, Gay Community Social Services, Seattle Counseling Services for Sexual Minorities, the Seattle Lesbian and Gay Community Center, and the Lambert House for queer youth have all been located on Capitol Hill, some for almost thirty years. The Lesbian Mother's National Defense Fund started in a bar on the Hill in the mid-1970s. Rallies and marches for Pride have gone down Broadway (the Hill's main thoroughfare) for decades. In the 1990s a Q Patrol was formed to combat homophobic street violence along Broadway by establishing a strong, visible queer presence on Capitol Hill. The 43rd State Legislative District, which contains both the Hill and the neighboring University District, is represented by two gay men, State Senator Ed Murray and State Representative Jamie Pedersen, and had previously been represented by Washington State's first openly gay legislator, Cal Anderson, in the 1980s.

Capitol Hill is also queer *culturally*. Gay and lesbian cultural expressions and meanings are inscribed into the landscape in multiple ways that show Capitol Hill means GLBTQ in Seattle. We see rainbow flags, the universal symbol of unity across difference for queer folk (fig. 6.51). We see same-sex couples holding hands and showing affection to each other in public spaces like sidewalks and open-air cafes. The Hill's parks are inscribed with queer culture as well: Cal Anderson Park is named for the legislator who died of AIDS in 1995. And perhaps more obliquely, Volunteer Park has been well known as an after-hours cruising area for men who have been having sex with men for decades.

Yet, saying a neighborhood represents or embodies a particular social geography—however true that is—is fraught with conditions, exceptions, qualifications, provisos, and ignorance. Indeed, the challenges are precisely what Queer Theory offers us: a will to challenge and destabilize (to make "strange") taken-for-granted ideas.[38] In this way, queer is a transitive verb rather than an adjective. To queer our notions of a gay neighborhood, we must consider that sexuality is not fixed in space. To say that Capitol Hill is the queer neighborhood in Seattle is not to say that it always has been so. It is not to say that other parts of Seattle are not—or have not also been—quite queer. It is not to say that there aren't important and different social geographies of gender that make collapsing men and women into "queer" districts difficult. It is also not to say that Capitol Hill itself is a fixed entity. On the contrary, it is always changing, as is the geography of the GLBTQ community. We want to suggest that it is helpful to consider the historical geography of queer space in Seattle to appreciate that it has always been *moving*. By "moving," we don't mean relocating per se; we mean that it has changed shape, scale, and direction. We mean that it changes depending on what is mapped and when it is mapped. We mean that it is helpful to think of queer space as being in flux. None of this is to deny that Capitol Hill is a gay space in the city, but it is more to challenge the assumption that relations between space and identity are fixed, when, in fact, they are more movable and changeable than our representations of them often admit. In this way, we follow a long-standing trend in social geography

6.51 GLBTQ sexuality reflected in the landscape of Capitol Hill.

6.52 Sign at the door of Pony, a Capitol Hill gay bar, 2010.

that takes a very familiar and ordinary urban form and shows a more complex, fluid, and ironic geography.

Looking at figure 6.53, which is a map of major sites of significance for the GLBTQ community since the 1930s, we can see the importance of Capitol Hill, though we might be surprised to admit that the Pike-Pine corridor (adjacent to but not technically "on" the Hill) has had more GLBTQ sites than Broadway, or that what was once called "Renton Hill"

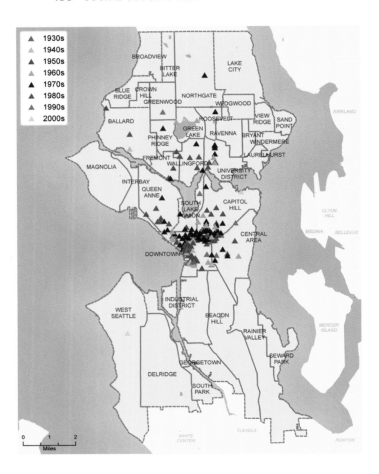

▲	1930s
▲	1940s
▲	1950s
▲	1960s
▲	1970s
▲	1980s
▲	1990s
▲	2000s

6.53 Major sites of significance for queer Seattle during the twentieth century: gay establishments by decades.

(the higher part of Capitol Hill east and south of Broadway, around Fifteenth Avenue) also seems to have been geographically important. We can also see clusters of sites in Pioneer Square, the downtown area, the University District, and Wallingford, and even around Queen Anne Hill.

Historically speaking, Capitol Hill is a relatively recent site for GLBTQ life. Throughout most of the twentieth century, that life was centered most visibly in Pioneer Square. The earliest recorded bar frequented by "homosexuals" was the Double Header, at the corner of Second and Washington. This venue is still open in 2010, still advertises in the *Seattle Gay News*, and features historic photos and pictures from "back in the day." Around the corner from it,

the Casino was an after-hours bar that opened after Prohibition (later it became a regular-hours spot), and from the 1940s to the 1960s, it was one of the very few spaces where same-sex dancing was allowed. It was fondly called "Madam Peabody's Dancing Academy for Young Ladies." From the 1930s through the late 1970s, a spate of bars, taverns, restaurants, and bathhouses were known to be queer, including (but not limited to) the Columbus Tavern, the Mocambo, the South End Steam Baths, Sappho's, the Grand Union, the 611, the 614, the 922, the Caper Club, the Golden Horseshoe, the Silver Slipper, the Submarine Room, the Blue Banjo, and Shelly's Leg. Of course, Pioneer Square was never *just* a gay space, and it was never known as a gay neighborhood per se. Homosexuals and other marginalized groups in society (single, working-class men; sex workers; people of color, etc.) congregated below the "deadline" (Yesler Way), away from the middle-class respectable social geography of Seattle. These different identities and communities were often segregated from each other at the micro-scale or used the same spaces (streets corners, alleys, stores, taverns) uncomfortably or with diffidence.

Complicating the story further, one cannot simply say that the gay district "moved" *from* Pioneer Square *to* Capitol Hill, because there was never a residential concentration of GLBTQ who moved from one place to the other, and because other parts of the city also provided queer space important to community formation and identity. Take the downtown core, which has historically held several important sites for gays and lesbians. Many of these included places that were not necessarily "gay," but where gay people and activities were tolerated (to a point). For example, in the 1950s and 1960s, the Marine Room in the basement of the tony Olympic Hotel was a notorious bar where young men, some of them hustlers, could pick up men. Similar stories are told about the old Ben Paris restaurant on Pine Street from the 1950s. In terms of bars and taverns, the Madison at 933 Third Avenue was a women's bar in the 1950s and 1960s. It became one of Seattle's first leather bars, the 922, in the 1970s and ended its existence as the lesbian Riverboat in the 1980s. The Pines, a bath-house, stood where Convention Center Station is now, next to Spags (1969–87), Disco Seattle (c. 1977), and Mike's (a gay tavern in the 1970s). The Lesbians of Color Caucus had an office in the 1300 block of Third Avenue and offered counseling, emergency assistance, and rap sessions in the 1970s. Probably the most famous cabaret was the Garden of Allah, from 1946 through 1956 located on First Avenue and University (roughly where Harbor Steps are now).[39] Here, legendary performers like Paris Delair, Jackie Starr, and Hotcha Hinton appeared to a largely mixed crowd—but in a venue that was decidedly queer, not to mention gay-owned (Fig 6.54).

Wallingford and the University District were also vital centers of gay life, especially for the lesbian community, in the 1970s. Collective houses like the Hespera House on Thackery Avene and Dyke Corner on First Avenue NE were important residential spaces where lesbians could create safe and feminist domestic spheres. The Gay Women's Resource Center was at 4224 University Way NE, in the heart of "The Ave." And the Lesbian Resource Center began in the University District YWCA in 1971 and had several locations, including on Roosevelt Way. Places such as the It's About Time Bookstore, Arabesque, and the Innerspace

Coffee House were retail environments where lesbian community and identity were nurtured and formed. Gay men also found space in these areas. The Bus Stop bar on North 45th Street began in 1987 and is now known as "Changes." And down by the I-5 overpass, Freeway Hall was a popular venue for gay and lesbian dances.

Other neighborhoods also claimed queer space here and there over time. Queen Anne Hill's Kinnear Park was known as a "beat" since at least the 1950s, and had several well-known and beloved houses shared by gay men, including the Kinnear Hilton on West Mercer Street, the Lollie Parkins House on Olympic Place, and the Gayler Palace (on West Galer Street). West Seattle had a short-lived gay bar called "Guppies" on the California Avenue business strip (next to the West Seattle Chamber of Commerce) in 2001–4. And the Central District has been home to the Gay Community Center, the Seattle Men's Chorus, and POCAN (People of Color Against AIDS), among other community organizations. The Gay Community Center was even briefly housed in the Seattle Urban League building there. And there was a short-lived (c. 1979) gay disco in the nearly suburban Northgate area (it was called "Karma").

The gay ghetto seems to move around if you consider

6.54 "Tame that Tiger" review at the Garden of Allah, c. 1953. Left to right: Billy DeVoe, Jackie Starr, Hotcha Hinton, Frances Blair, and the MC, Kenny Bee.

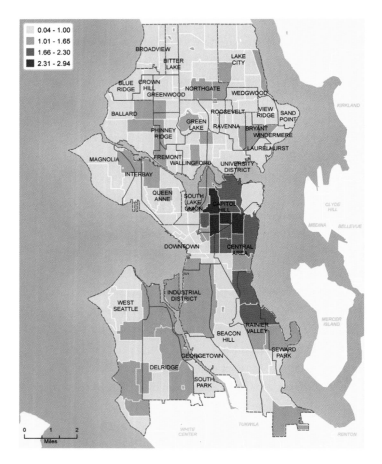

6.55 Location quotients of all same-sex households, 2000. Source: U.S. Census.

just what needs to cluster in order for a district to be considered "gay." The 2000 Census was the first one in which same-sex households could be seen nationally. Figure 6.55 shows the map of same-sex households by census tract in Seattle, using location quotients (recall from earlier in this chapter that a score of 1.0 means that that tract has less than its share of couples relative to the city overall. A value greater than 1.0 means that tract has that many times more than its share of same sex couples). Here we can see that while the Capitol Hill tracts surely do exhibit the highest proportions of same-sex households, other neighborhoods not typically thought of as gay also show a high concentration: the Central District, Madison Valley, Madrona, Leschi,

and even Mount Baker and Seward Park as well.

Gender also moves the ghetto around. If we compare proportions of male couples with female couples, we can see rather different social geographies in Seattle (compare figs. 6.56a and 6.56b). Although they do overlap in some census tracts, lesbian couples have a much more diffuse residential geography than gay male couples. This no doubt reflects the wage and class differentials between women and men that intersect with Seattle's housing market. These maps are based only on queer folks who self-reported their domestic partnerships on the census form. If we could map all of queer Seattle, the patterns would undoubtedly be even more diffuse.

Historically, it was especially difficult to secure housing if you were an out gay or lesbian person, because of discrimination that was legal until the 1970s. Apartments and rooming houses became important means for queers to claim home in residential space. And before there was a "gay ghetto," such places, albeit discreet, could be found all over the city. During the 1940s, the Governor Hotel at Spring and Fifth was a residential hotel popular with homosexuals. Another popular residential hotel for gay men was in Ballard at the Starlight Hotel. In the 1950s, the Hotel Richelieu at Union and Sixth Avenue was known to be friendly toward gay men. In Leschi, the Alder House was a well-known gay men's group house from 1957 to 1966. Skippy La Rue was a legendary gay performer in the 1950s and 1960s. He owned a couple of houses (in Fremont and Capitol Hill) throughout his time in Seattle, and they were always safe places for gay men to rent rooms. In the 1970s, lesbians also formed shared housing arrangements in typically gay-male parts of Capitol Hill, such as the "Red Hen" at John and Eleventh and "Pud Street" on East Mercer Street.

Gay and lesbian Seattleites also claimed space beyond Seattle. Since the late 1970s, there have been gay-friendly campgrounds, first at Index, then near Granite Falls (both northeast of Seattle in the Cascade Mountains). In the 1970s, a back-to-the-land group within the gay community

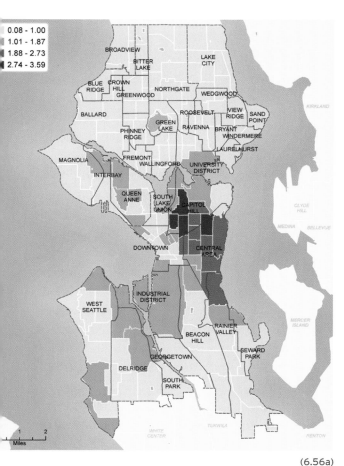

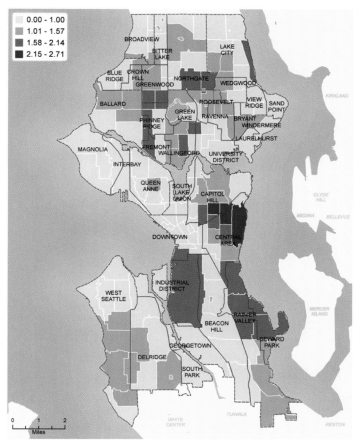

(6.56a)   (6.56b)

6.56 Location quotients of same-sex households, 2000: (a) male couples only; (b) female couples only. Source: U.S. Census.

purchased land at Elwah on the Olympic Peninsula. Proceeds from the sale of that land were used to form Gay Community Social Services, one of the first organizations of its kind. Vashon Island and North Bend, exurban rural areas in the region, also exhibited high proportions of same-sex couples (fig. 6.57). There have been gay bars in both Everett and Kent for decades. Interestingly, there are no records of bars on the Eastside (where Bellevue, the state's fifth largest city, is located).

It is often said that "gays are everywhere" when discussing the geography of queer folk. And indeed that's true.

But the question is how? What we've described so far are exceptions to Capitol Hill as the only GLBTQ area in Seattle. We write this at a time when there is growing acceptance of GLBTQ people in the United States. Washington State now has domestic partnership legislation that is informally known as "everything but marriage." At the same time, the number of gay and lesbian bars in Seattle is on the decline (similar to trends in other U.S. cities), and the Pride Parade has moved downtown to Fourth Avene from Capitol Hill, to gain more space and greater integration into the city culture. Most Seattle neighborhoods have GLBTQ residents—either singles or couples. Does this movement, with all its ubiquity, still need Capitol Hill, a sort of psycho-geographic

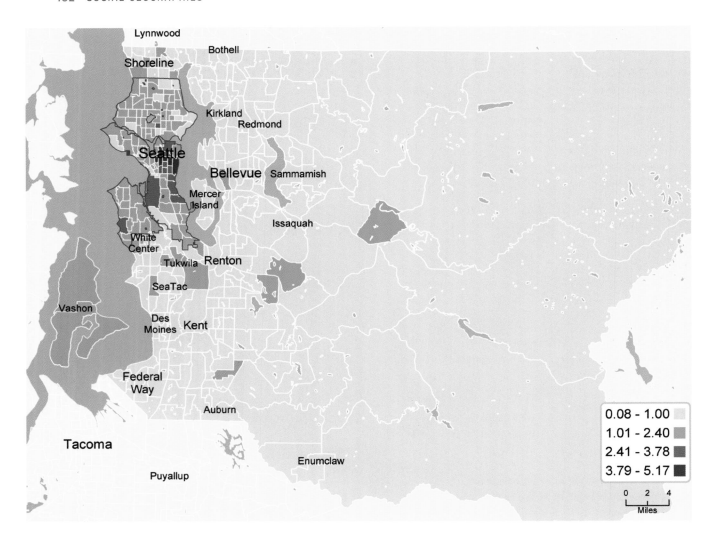

6.57 Same-sex couples across King County, 2000.
Source: U.S. Census.

anchor (however anachronistic), to tether all this diffu-
sion? If that is the case, it becomes even more important to
"queer" queer space, as new and multiple ways of claiming
space will evolve alongside the urban "gay ghetto."

NOTES

1   ForecastChart.com, based on the Federal Housing
     Finance Agency, Seattle Home Price Index.

2   Mortgage delinquencies are loans with payments that
     are at least 30 days late. As a first step toward fore-
     closure, a Notice of Trustee Sale is filed with the King
     County assessor's office prior to selling a mortgage-
     delinquent home at auction. Washington law allows
     both judicial and nonjudicial foreclosures. Generally,
     the property must not be sold within 190 days of the
     filing date of the notice. The borrower has until 11 days
     prior to the sale to pay off the delinquent amount, plus
     expenses and fees. Other options to forestall foreclo-
     sure can be pursued as well.

3   First, we gathered data on potential foreclosures to
     cover the period from January 2005 to August 2009.

The Notice of Trustee sales records are available to the public at the King County assessor's office, and numerous commercial Web sites, such as Prospera Real Estate, provide weekly listings of all notices. These data were provided by Susan Kinne (UW). Matt Townley provided the summary parcel information for the GIS data merge. Not all trustee sale properties will go to auction, necessarily, but these are the properties at risk. Some properties might proceed to short sale, and others may be purchased from the bank.

Second, 2009 demographic data were downloaded from the Federal Financial Institutions Examination Council (FFIEC). Third, King County parcel information was merged with census tract boundaries, enabling us to collect all potential foreclosure information for any census tracts. We could analyze data only at the census-tract scale; we were not able to analyze individual foreclosures because personal information is not included in the Notice of Trustee Sale. The Home Mortgage Disclosure Act (HMDA) is a source of publicly available data on mortgage demand and origination that provides information about the demographics of the borrower and the features of the loan. However, it contains no information about what happens to the loan after origination. Consequently, our analysis focused on the location of vulnerable properties rather than on the identity of vulnerable homeowners.

4   Data for this analysis were taken from "Population of Cities and Towns," Puget Sound Trends, No. D3, Puget Sound Regional Council, September 2009. "Housing Prices and Affordability," Puget Sound Trends, No. E16, Puget Sound Regional Council, August 2009. "Report to Congress on the Root Causes of the Foreclosure Crisis," U.S. Department of Housing and Urban Development, Office of Policy Development and Research, January 2010. "Delinquencies, Foreclosure Starts Fall in Latest MBA National Delinquency Survey," Mortgage Bankers Association, February 19, 2010. Federal Housing Finance Agency House Price Index, http://www.forecast-chart.com/prices-home-seattle.html.

5   Tent City 4 was created in 2004 to serve the homeless population of Seattle's affluent Eastside.

6   Seattle King County Coalition on Homelessness, "2009 One Night Count Summary," SKCCH, http://www.homelessinfo.org/downloads/2009_SKCCH_ONC_Summary_Numbers.pdf.

7   "A Roof Over Every Bed," Committee to End Homelessness in King County (Seattle: King County, 2005).

8   Parents Involved in Community Schools v. Seattle School District No. 1. 551 U.S. 701 (2007).

9   "The Resegregation of Seattle's Schools: A Seattle Times Special Report," Seattle Times, June 6, 2008; http://seattletimes.nwsource.com/flatpages/local/schoolresegregation.html.

10   Using the online data tool from the Seattle Times Web site, we have charted the changes in percent minority by school clusters for 1971, 1980, 1990, and 2007.

11   "Historical County Population Change, 1950–2000," Puget Sound Regional Council, 2001; http://www.psrc.org/assets/791/d1jul01.pdf.

12   City of Seattle Department of Design, Construction, and Land Use, "Birthplace of Seattle's Foreign Born Population," Demographic Snapshots, 2003, http://www.seattle.gov/dpd/cms/groups/pan/@pan/documents/web_informational/dpds_006762.pdf.

13   "A Tale of Two Seattles," Seattle Times, January 2, 2007; http://seattletimes.nwsource.com/ABPub/2007/01/02/2003506141.pdf.

14   "Historical Enrollment Information," Seattle Public School, 2010; http://www.seattleschools.org/area/eso/OntimeAnalysis_2009-10_Short.pdf.

15   Catherine Veninga, "Road Scholars: School Busing and the Politics of Integration in Seattle," PhD diss., University of Washington.

16   Thanks to Joe Eckert, who provided sterling research assistance with the Census data analysis that informs this essay, and to Ashley Hards, who fact-checked locations and buildings for me.

17   Jane Jacobs, The Death and Life of Great American Cities (New York: Random House, 2005).

18   See Pike Place Market Centennial, http://www.seattle.gov/CityArchives/Exhibits/PPM.

19   J. Walljasper and D. l. Kraker, "The 15 Hippest Places to Live: The Coolest Neighborhoods in America and Canada," *Utne Reader*, Nov./Dec. 1997.

20   http://www.belltown.org.

21   Jennifer Wolch and Michael Dear, *Landscapes of Despair* (Princeton, NJ: Princeton University Press, 1987).

22   Coll Thrush, *Native Seattle* (Seattle: University of Washington Press, 2007).

23   Ibid., 93.

24   Cited in M. Klingle, *Emerald City* (New Haven, CT: Yale University Press, 2007. Original source is *Clarence B. Bagley History of Seattle from the Earliest Settlement to the Present Time*, vol. 1 (Chicago: S. J. Clarke Publishing Co., 1916).

25   Ibid., 111.

26   C. Schmid, *Social Trends in Seattle* (Seattle: University of Washington Press, 1944). Schmid categorized the northern edge of Denny Triangle as "automobile sales and service," with "mostly business" immediately north of the area he marked as "Belltown." First Avenue and Pike Place Market were described as filled with "mostly cheap hotels, second hand stores and taverns"; between there and the waterfront was an "industrial, warehouse and wholesale area."

27   Mikala Woodward, *Rainier Valley Food Stories Cookbook* (Seattle: Rainier Valley Historical Society, 2005).

28   Ibid. See also Cassandra Tate, "Seattle Neighborhoods: Columbia City Thumbnail History," http://www.historylink.org/index.cfm?DisplayPage=output.cfm&File_Id=3327.

29   Trevor Griffey, "Preservation and Economics in North Rainier Valley, http://www.historicseattle.org/preservationseattle/neighborhoods/defaultoct2.htm.

30   U.S. Bureau of the Census, *1970 Decennial Census: Selected Demographic, Socioeconomic and Housing Data for King County Census Tract 103 and Seattle City; 1980 Decennial Census: Selected Demographic, Socio-economic and Housing Data for King County Census Tract 103 and Seattle City; 1990 Decennial Census: Selected Demographic, Socioeconomic and Housing Data for King County Census Tract 103 and Seattle City* (Washington, DC: U.S. Bureau of the Census, 1970, 1980, 1990), Summary Tape File 3 in factfinder.census.gov (accessed June 15, 2008).

31   U.S. Bureau of the Census, *1980 Decennial Census* and *1990 Decennial Census*.

32   U.S. Bureau of the Census, *1990 Decennial Census,* Tape File 3, in factfinder.census.gov (accessed June 15, 2008).

33   Seattle Planning Commission, "Seattle's Neighborhood Planning Program, 1995–1999," 1999, http://www.cityofseattle.net/planningcommission/docs/finalreport.pdf.

34   Seattle Planning Commission, "Neighborhood Plan Stewardship Survey: A Snapshot of Plan Stewardship in Seattle," no date, http://www.seattle.gov/planningcommission/docs/stewardshipsurvey.pdf.

35   Based on informal conversations and interviews with residents.

36   All respondents had to have lived in the neighborhood for at least three years and be at least twenty-one years old. Respondents were classified as "stayers" if they had lived in the neighborhood for at least ten years and had earned under $40,000 the previous year, and as "gentrifiers," if they had lived in the neighborhood for under ten years and earned over $60,000 the previous year.

37   Gary Atkins, *Gay Seattle* (Seattle: University of Washington Press, 2003).

38   Ann Marie Jargose, *Queer Theory* (New York: NYU Press, 1996).

39   Don Paulson and Roger Simpson, *An Evening at the Garden of Allah* (New York: Columbia University Press, 1996).

SEVEN

CULTURAL GEOGRAPHIES

Katharyne Mitchell and the 2010 UW Geography
Undergraduate Honors Students:
Mikail Aydyn Blyth, Ethan Boyles, Sofia Gogic,
J. E. Kramak, Rita B. Lee, Hayley Pickus,
George Roth, Anne Steinberg, Nicole S. Straub,
Lola S. Stronach, and Carl Urness

Cultural geography is a difficult field to pin down. It generally refers to a study of the cultural forms that are evident in the landscape, and the meanings and practices that produce them and are ascribed to them. These landscapes are made meaningful in countless ways, often reflecting dominant interests in society, but are also frequently challenged by those who feel excluded. What is emphasized, what is ignored, who is able to produce and represent landscapes and who isn't are questions that reflect style and the currents of fashion, but are also the power relations of a given time and place. Investigating these latter issues is a critical focus of cultural geographers, who try to analyze what is absent as well as what is present in cultural productions, and who also examine the multiple underlying structures that condition the form and expression of place.

Landscape production in Seattle often manifests a tension between a deep connection with the natural environment and the desire or necessity to make a strong human imprint on the landscape. It also reflects tensions between those who seek to profit from the land and those who want to preserve the past or manage resources in more equitable or sustainable ways. Many different kinds of struggles over equity and the future of the city are played out literally in the streets—for example, the massive WTO protests in Seattle in 1999—but are also evident in public artworks, neighborhood aesthetics, urban gardens, music, and many other forms of cultural expression. In what follows, we provide something of an insider's guide to a few of the well-known cultural icons of the city and also introduce some of the more subtle practices and expressions of art and life in Seattle in the twenty-first century.

The Salmon: Our (Conflicted) Heart

It's a rainy November day—not unusual for Seattle—and throngs of tourists and locals alike are huddled inside Pike Place Market waiting for the show. What they're waiting for is not any of the buskers, such as the woman who clog dances on a washboard with an old-time string band for backup or the guy wandering around with the live opossum on his shoulder. It's the fish. They're waiting for the mongers at the Pike Place Fish Market to perform that oft-repeated ritual of fish throwing.

When the order comes for a whole salmon, one of the workers plucks the chosen catch from its icy bed, slinging it fifteen feet over the counter to his colleague, who deftly catches the slimy missile, wraps it in paper, and presents it to the customer. It's hard to tell who the star of this show is, the thrower or the catcher. Or maybe it's the fish itself? That singular image of living silver, in both its flying and swimming forms, is not just a cultural icon of the Emerald City; it's also a source of conflict. The salmon is deeply embedded in the past, present, and future of Seattle and is etched indelibly into its living culture (fig. 7.1).

The salmon has been a cornerstone of Pacific Northwest culture since long before European settlers arrived. It figures prominently in the religion and mythology of the indigenous peoples of the region, and it was of inestimable importance as both a source of food and as a spiritual being. With European colonization and the forced social, political, and cultural institutions that came with it, there was a distinct understanding among Native Americans living in the Seattle area that their lives and their relationship with the salmon would change. The enterprising leader of the Suquamish and Duwamish tribes, Chief Si'ahl (Seattle's namesake), hoped his people's relationship with the new settlers would provide for the prosperity of all. After the establishment of the Washington Territory in 1853, Si'ahl helped secure Native rights to the salmon of Seattle's rivers and the Puget Sound. Despite having ceded enormous amounts of land through treaties with Washington Governor Isaac Stevens in 1854 and 1855, Chief Si'ahl maintained

Native Americans' rights to fish for salmon. These treaties have played an ongoing role in historical and contemporary clashes between Native and nonnative Americans in Seattle over this valuable resource.

As the nonnative fishing interests in the central and southern reaches of Puget Sound grew, treaty tribes and their traditional and semi-traditional fishing methods were gradually pushed to the periphery. Euro-Americans' efficient fishing techniques, combined with state regulations on the salmon industry, effectively denied Native Americans access that had been guaranteed to them by the treaties of the 1850s. There was continuous contention over salmon fishing rights in the Seattle area throughout the twentieth century. By the 1940s and 1950s, questionable enforcement of fishing regulations limited Native access to the salmon harvest to reservation waters; those who fished

7.1 Salmon at the Pike Place Market.

outside these waters risked confiscation of equipment and/or imprisonment. Numerous legal battles took place between treaty tribes and the state, resulting in often contradictory judicial rulings.[1]

In 1963 the *Washington v. McCoy* decision acknowledged that the state was entitled to exercise regulatory control over Native salmon fishing for the purposes of conservation, seemingly in contradiction of the treaties. *Washington v. McCoy* fueled the flames of discord and precipitated the formation of a group committed to the protection of Native fishing rights through civil disobedience, called Survival of the American Indian Association (SAIA).[2] Inspired by the civil rights protests of African Americans at the time, SAIA and the National Indian Youth Council (NIYC), in conjunction with local tribal members, staged a series of "fish-ins." During the fish-ins, tribal fishermen would harvest salmon outside of their reservation waters without state-issued permits, directly confronting law enforcement officials. Protests by Native Americans over fishing rights were championed by luminaries like Marlon Brando, who participated and was subsequently arrested during a fish-in in 1964. On the other side of the struggle, sport and commercial fishermen reviled the protests as exercises of unacceptable privilege. In January of 1971, a Native fisherman was shot in his sleep while tending nets on the Puyallup River. The alleged perpetrators were two white sport fishermen.

Just a year prior to this violence, the federal government had filed suit against Washington State, relating to the state's regulation of Native fishing practices. Responsibility for the case was given to Federal District Judge George Boldt. After spending more than three years researching the case, Boldt issued his decision: *United States v. Washington*, better known as the Boldt Decision, determined that local tribes were entitled to 50 percent of the harvestable salmon in "usual and accustomed fishing areas of the treaty tribes." They were now free to fish salmon virtually uninhibited by state regulation outside of reservation lands.

Judge Boldt's decision did little to alleviate the dispute. The value of this cultural and economic icon far outweighed the court's ruling, and the conflict continued, pitting less than 1,000 treaty Indians who made their livings fishing salmon against nearly 6,600 nontreaty commercial fishermen and more than 280,000 licensed sports fishermen. With so much at stake, commercial and sport fishing interests continued to battle the ruling to the extent that five years later, the U.S. Supreme Court noted that "except for some desegregation cases, the district court faced the most concerted official and private efforts to frustrate a decree of a federal court witnessed in this century."[3]

Recently a new player has entered the fray, mitigating the historical animosity and uniting the two sides: the haunting possibility of species extinction. In March of 1999, NOAA's National Marine Fisheries Service listed the Chinook salmon as a threatened species. A combination of overfishing, habitat destruction, and pollution has placed this vital resource in such a precarious position. An alternate future could now be imagined—one without the Chinook salmon.

Urban pollution has been a major contributor to the salmon's decline. In an almost immediate response to the listing of the Chinook as threatened, the city started the Salmon Friendly Seattle (SFS) program. This program is designed as a continuing research project, a source of information for citizens, and a call to action to help preserve the salmon. From salmon friendly gardening techniques to volunteer shoreline restoration projects and to cultural art and informational brochures, SFS has helped to embed the importance of salmon into the collective imagination of the city.

A brief tour of Seattle and the surrounding area reveals images of the salmon in the airport, on highway overpasses, on sidewalks, at bus stops, and stenciled next to storm drains, reminding us of how entwined our daily lives are with nature. From historical conflicts (with important cultural implications) to modern conservation efforts, the importance of salmon to the Puget Sound region cannot be overstated. As an icon of multiple cultures within the area, there is no doubt that this symbol will continue to shape and reveal the trajectory of interac-

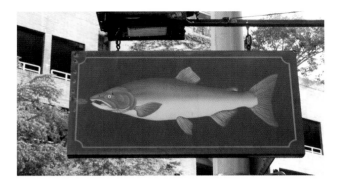

7.2 Salmon sign.

tions between the diverse populations of the Seattle area and the natural environment. As with Seattle's relationship with its parks, the city has worked at reimagining the postindustrial city as a place in which residents act in harmony to protect the salmon.

From Industrial to Postindustrial City

Boeing is a company name that has always been attached to Seattle, and it is impossible to speak of the culture of Seattle without addressing the critical role of this corporation. After the timber industry began to decline, Boeing stepped in to fill the industrial hole, and its regional influence continued to grow (see chapter 2). By the 1950s, Boeing was becoming one of the most important companies in the transportation industry because of the increase in air travel, just as transportation was becoming *the* industry in Seattle. So it was in the interests of city and state officials to keep Boeing grounded in Seattle. The company was provided with tax breaks, subsidies, and lucrative contracts, both at the national and municipal levels.

In September 1968, Boeing delivered its latest aeronautic tour de force: the massive 747 jumbo jet, the largest passenger airplane of its time. The thousands of engineers and assembly workers who worked on the project came to be known as "the Incredibles," and the images of these laborers and of the majestic plane itself became deeply etched into Seattle's culture. Boeing's dominance fulfilled

the lofty promises laid out by the World's Fair of 1962, and both Boeing managers and city officials looked forward to a prosperous future together. But, as chapter 2 has described, the boom of the 1960s gave way to the bust of 1969. Many of the Incredibles found their work disappearing, as employment at Boeing dropped from over 100,000 jobs in 1968 to less than 40,000 by April of 1971, a two-thirds reduction in the work force.

As the saying goes, when Boeing sneezes, Seattle catches a cold. As a result of the Boeing Bust, Seattle unemployment rose from 2.5 percent in 1968 to an astounding 17 percent in 1971, the highest in the country and more than twice the national average. Economic reports indicated that it was one of the sharpest rises in unemployment of any major city since the Depression of the 1930s.

The cultural malaise of the city was effectively summed up in that billboard that asked, "Will the last person leaving Seattle please turn out the lights?" Thousands of laid-off employees saw a dim future in Seattle, and they left the city in search of clearer skies. The *Seattle Times*'s classified section was flooded with ads for homes and cars for sale at rock-bottom prices.

Ultimately, this moment of deindustrialization also signified a moment of change and rebirth. Seattle's reputa-

tion for being an innovative and evolving city, paired with its historical links to the lush nature surrounding it, set the stage for urban reinvention. Movements to rework certain aspects of the city were propelled by redevelopment and gentrification, often at the cost of both history and people. At the same time, there were also strong grassroots efforts made to preserve the historical Pike Place Market and Pioneer Square, as well as Seattle's unique houseboats and houseboat culture (made nationally famous by the film *Sleepless in Seattle*), and to protect vulnerable populations and recycle and remember abandoned industrial neighborhoods.

One of the most striking testaments to this moment of economic transition (and the controversies surrounding it) is Gas Works Park, an area converted from a former gasification plant on the northern shore of Lake Union (fig. 7.3). The land, formerly known as Brown's Point, was purchased by the Seattle Gas Light Company in 1906 and became thereafter a site for manufacturing the gas that supplied the lamps and heaters of the surrounding homes and businesses. It was active for the first half of the last century, from 1906 to 1956, mirroring the rise of Seattle's major industrial period. Shortly following the Boeing crash, and after nearly twenty years of dormancy, the area was reopened in 1975 as a recreational park.

The story of this park is a story of cultural struggle over Seattle's past, present, and future. The city acquired the twenty-acre site in 1962, after a major effort by Councilwoman Myrtle Edwards, who was chair of the Parks Committee. After funding was secured in 1968, the landscape architect Richard Haag was awarded the commission to remake the site. Haag was determined to keep many of the old industrial buildings intact on the site as a memorial to a bygone era. In addition to exposing the giant factory pipes and technologies for public view, he also recycled materials from the plant, such as the thousands of cubic tons of rubble that are now piled under fresh soil, to create the "Great Mound," a kite-flying hill overlooking downtown Seattle. The design proved highly controversial, with Myrtle Edwards's family refusing to give the park her name. But Haag felt it conveyed Seattle's transition from industrial to postindustrial city, and in the end, the design was accepted unanimously by the city council. Haag's design was part of a broader movement to popularize the city's shift away from its manufacturing base, with the rusting industrial buildings indicating a bygone era that would soon be transcended by a shining, new service-oriented metropolis.

7.3 Gas Works Park.

In the words of one national commentator:

Gas Works went from being the place of industrial urban rejection to the premier site for the aestheticization of the new urban landscape of a service-oriented office culture and the valorization of a local "postindustrial" narrative. It was the space on which Seattle's economy pivoted between 1968 and 1971. Historic preservation of the Gas Works accomplished much of the groundwork necessary for the social production of contemporary Seattle.[4]

Perhaps such a direct physical manifestation of Seattle's economic transformation is one reason why Gas Works Park remains a popular location for political rallies and social events. For example, the Summer Solstice Parade and Pageant, an alternative, bohemian celebration of a pagan holiday, designates the park as its endpoint. Naked bicyclists, topless carnival dancers, giant colorful puppets, and people dressed in unusual attire convene on the recycled terrain to commemorate the mixing of the old and the new, of memory and fantasy, in the spirit of Seattle. The park has also been a key political rallying point, serving as the site for a seven-month Seattle vigil against the first Gulf War in 1990. Named Peace Works Park by those at the vigil, the location, history of struggle in its creation, and the unusual artifacts present on the site, all made it a central and iconic platform from which to express political dissent.

Gas Works Park was one of many reworkings of the urban landscape that occurred between the 1968 bond measure Forward Thrust and the 1991 Growth Management Act. These projects involved a dramatic re-envisioning of the city, interlacing economic growth with a new "postindustrial" culture that urban elites in the city hoped to perpetuate. Indeed, despite the use of the park for political rallies, a more critical image of this refashioned space is not one of urban renaissance but rather of a "bourgeois playground."

Postindustrial, neoliberal city politics, which turned away from welfare provision toward a more entrepreneurial positioning of Seattle, can also be detected in Forward Thrust. Begun in 1966 by attorney James Ellis, Forward Thrust proposed a $333.9 million coordinated program of capital improvements, including the major projects of a basic rapid transit system; a domed, major league sports stadium; major arterial street improvements; low-income housing; sufficient parks, plazas, and greenbelts to satisfy metropolitan needs; a world trade center; and the permanent elimination of urban sprawl.

Only the rapid transit system and low-income housing did not pass. The approved Forward Thrust projects dramatically reshaped the city landscape and culture, creating an image of a high quality of life for a postindustrial, urban elite (through the creation of world-class parks, a world trade center, and sports facilities), while neglecting the needs of the poor. James Ellis himself noted this glaring omission in a *Seattle Post-Intelligencer* editorial: "Why, when we are cleaning up our waters . . . and cultural activity and recreational facilities are in a virtual renaissance, [are we] failing to provide the good schools and safe streets that make urban living civilized?"

As Seattle's economy shifted toward service-sector employment, its skyline began to change rapidly. While most of the downtown office buildings constructed in the 1970s and 1980s reflected the soulless corporate spirit of that era, a later architectural renaissance saved the city from complete design oblivion. The unique and controversial Experience Music Project (EMP) building, commissioned by software billionaire Paul Allen and designed by Frank Gehry, was one of the first major buildings to challenge the boxy, dark-glass image of the previous decades. Created to echo the lines of Jimi Hendrix's busted guitar (or electric guitar trash in general), the building used bright color and bent form to evoke strong reactions—similar to the early feelings experienced with the advent of rock-'n'-roll (see fig. 3.4).

Another recent icon that manifests Seattle's newly acquired "cultural sophistication" is the Seattle Central Library, completed in 2004 (fig. 7.4). The downtown public library has gained international attention as one of the most architecturally advanced buildings in the world. The asymmetrical design is bold and unusual, and the building serves its purpose as an extensive and fully functional library, while also incorporating public spaces in the interior. Unlike the inward-oriented buildings of the 1980s, the library helps to achieve and maintain Seattle's image as a natural or environmental city by featuring public spaces that have extraordinary views of Elliott Bay. While Seattle gained recognition as a technologically advanced city with the dotcom boom in the 1990s, the completion of the library helped it to maintain this status even as the ailing dotcom industry declined.

As Seattle's skyline expanded, the former president of the Downtown Seattle Association told the *Seattle Times* that "the glittering new crop of downtown office towers symbolizes vitality, progress. I look at it and see beauty." At the time, the changes in downtown Seattle were preparing the city to be a global center of trade and finance, which included all the right ingredients: professional ball clubs, sushi bars, espresso stands, a convention center, sculpture parks, and skyscrapers. The city is beautiful. But there has also been a significant cost for this new skyline and its world-class image for many workers, struggling to find a place amid declining manufacturing jobs and rising prices.

The entrepreneurial city and all of its cultural symbols were never targeted for these workers, but instead were aimed at the producer services/high-tech, educated classes of a postindustrial Seattle.

Coffee and the Era of the "Microsofty"

In *Selling Seattle*, author James Lyons provides a snapshot of the release of Microsoft's Windows 2000. At the launch party, then CEO Bill Gates stood in front of a giant screen, replete with a logo containing the Space Needle glittering behind the nighttime city skyline. Lyons noted:

> The Space Needle, a slender, iconic tower constructed for Seattle's Century 21 Exposition in 1962 . . . was the centerpiece of Microsoft's shimmering simulated Seattle as a Windows 2000 logo whizzed around it to dizzying effect. For those watching, the inference was clear: Microsoft, Seattle, and the shape of the future were inseparably interfused.[5]

Microsoft's choice of the Space Needle might have seemed anachronistic (why hearken back to an earlier era

7.4 Seattle Public Library.

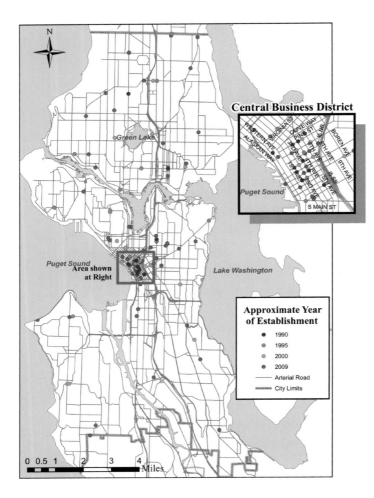

Central Business District

Approximate Year of Establishment

- 1990
- 1995
- 2000
- 2009
— Arterial Road
═══ City Limits

7.5 Starbuck's locations in Seattle.

three decades, as it has in many cities around the world. As evident in figure 7.5, its Seattle shops are located predominantly within the business district to attract customers able to afford the high cost of the gourmet beverages.

Starbucks has become a daily part of life for many Seattleites and a powerful influence on the cultural landscape of the city. As a self-proclaimed "third place" (alongside home and workplace), it provides a neutral, safe, public gathering spot between home and work. The company designed its stores with large windows and created many outward-facing seating areas and outdoor patios, thus allowing customers to "eat the street." This was a direct rejection of a type of 1980s development known as "killing the street," because of its inward orientation. Beginning with its first Seattle stores, Starbucks created a new, highly successful, corporate coffee culture that invited its customers to stay as long as they wanted and still feel like a part of the outside world.

The explosion of Starbucks locations worldwide, with their ubiquitous Starbucks lingo and signature products, produced other cultural expectations and norms as well. The availability of 87,000 different beverage combinations encouraged regular customers to customize their personal tastes and created an image of high social status, education, and sophistication that was tantamount to equating a $4.00 latte to an elite social identity. Because the majority of Starbucks stores' growth took place during the age of disposable (for some) income of the 1990s, Starbucks became a desirable luxury for a new group of consumers.

During the boom years of Starbucks's global expansion, however, downtown office space in Seattle was overdeveloped and vacancy rates increased. Additionally, the long-standing department store Frederick and Nelson went bankrupt and left its downtown office at the same time that Nordstrom (a mainstay Seattle-based company) was looking to move its downtown headquarters to the suburbs. Thus, at the same time that the seeds to remake the city into a high-tech, coffee-lover's paradise were being planted, a separate path that would have a substantial (and often ignored) impact on the cultural geography of Seattle

of promise?), but it brings home the point that Seattle's push for prominence is decades old, and that structures in the built environment are influential by their presence—or absence—in shaping the future visions of a city. While high-tech industries seem to rise out of nowhere to be the wave of the future, it is important to note that Microsoft might never have happened in the city without Boeing. The aerospace giant created a kind of educated, middle-class, tech-friendly community that would nurture Bill Gates and attract the workers Redmond needs to thrive. In other words, Boeing planted the seeds from which sprouted Seattle's saving grace: Silicon Forest.

Part of the culture of the new Forest is, of course, gourmet coffee. The first Starbucks store was established in Seattle's Pike Place Market in 1971, and the number of Starbucks in the city has grown exponentially over the last had been ongoing. Many of the Seattleites hit hardest by the Boeing bust and its aftermath were those who held low-wage, primarily manufacturing and assembly-line jobs. The boom in high-tech and service industries did little for this sector of the population and has led to increasing economic disparities in the city over time. As manufacturing's total impact on the economy went from 30 percent to 18 percent by 1980, those no longer within the industry would have to find work elsewhere in the city or leave.

Reconnecting to the Earth: Seattle's Alternative Food Movements

A recent push to forsake commercial food processing has fueled a new food culture in Seattle. It is built on a more direct connection to nature. Interconnectedness to the land provided Puget Sound residents with sustenance for many generations. In contemporary agri-business practices, however, the intimate relationship to the land and the seasonal aspects of farming have been lost with the increasing globalization of the food-growing process. Yet, sustainable food practices in the region have not been entirely lost. Farmer's markets, food cooperatives, and community gardens (P-Patches) have helped revitalize a more direct and personal connection with the nourishing aspects of the land.

Even beyond housing Seattle's famous fish throwers, Pike Place Market itself is a cultural icon for Seattle, representing the region's positive outlook toward fresh food and local farmers. Since 1907 the Market has been an outlet for local farmers to reach consumers without the interference of commission houses (fig. 7.6). The significance of Pike Place Market is profound, as it represents local desires to pay producers a fair wage for their farming labor and also to shrink the literal and symbolic distance between the rural and the urban. Fortunately, the cultural atmosphere of the market has been largely preserved, despite the threat of demolition in the 1960s. Much of the defense of the market is attributable to the vehement dedication of the Seattle community, who have protected it for over four decades.

7.6 Pike Place Market ca. 1912.

Now home to almost two hundred local farmers and craftspeople, the Market provides a face-to-face meeting place for food producers and Seattleites and tourists, even in an age of increasing distance between the dinner table and the farm. Named the "Soul of Seattle," Pike Place Market provides local Seattle flavor as well as culinary diversity. In the early 1980s, it became home to Hmong and Mien refugees from Laos, who found purpose and success in selling vegetables at market stalls and who subsequently reshaped local food by making Asian vegetables, such as bok choy, more widely available. Not only is Pike Place Market a venue for local farmers to earn decent wages, it represents one of Seattle's first pushes toward a more sustainable and local outlook on food.

Like Pike Place Market, Seattle's local food cooperatives are unique alternatives to contemporary food supplies. Consumers seeking a more healthy and local approach to food formed Seattle's renowned Puget Consumer's Cooperative (PCC) in the early 1950s. Founders of PCC were part of a broad counterculture that promoted coexistence with nature. PCC began as a grassroots movement, using newsletters to bring awareness of local and organic produce to the Seattle community. After twenty years of meetings at members' homes and community centers, PCC finally gained the cooperation and support of local growers and community members needed to open its first store in Seattle's Ravenna neighborhood.

Locally based markets were only the first step toward a new food culture. For the past fifty years, Seattle residents have used urban gardens as a means to reconnect with nature within the confines of the urban environment. The popularity of backyard and community gardens stemmed from the promotion of "Victory Gardens" in World War II as a way to increase the public supply of vegetables and herbs. These gardens became immensely popular during the war as well as shortly afterward, but their prevalence eventually declined. Urban gardening advocates, however, used the community network established by PCC to spread the news about backyard and community gardening and to distribute knowledge and tools to help others create their

7.7 The Picardo P-Patch.

own gardens. Over time, PCC became the local authority on growing food noncommercially and helped vault urban gardens into a cultural phenomena.

As interest in the practice grew, PCC member and University of Washington student Darlyn Rundberg Del Boca in 1973 received permission from the Picardo family to use part of their local truck farm for a community garden. This parcel of land was eventually purchased by the city with the help of several city council members and the mayor. Mayor Wes Uhlman, who had a garden plot himself, established an urban gardening program in Seattle that quickly spread to other parts of the city. The Picardo farm continued to be the model for new urban gardens, thus the new plots were nicknamed "P-Patches," a term that has become a part of Seattle's cultural vocabulary (fig. 7.7).

Grunge: Seattle's Alternative Music Scene

Few music scenes have fostered as much place-specific recognition as the Seattle grunge movement of the early 1990s. Plenty of scholars and music writers have tried to explain the phenomenon as something local and regional, tying it to Seattle's mix of landscape and isolation. But are these valid arguments? Is musical serendipity dependent on location alone, or was there something deeper at work in Seattle's garages?

Seattle's "sound" is the sort of intangible concept that most writers looked for as they examined Seattle's musical history for any hints of a single evolutionary process or catalyst to define what would become known as "grunge." But aside from a few hints from the "noisy" sound of Jimi Hendrix, nobody could point to any two Seattle bands between the 1960s and the late 1980s and say with credibility that they were following any direct lineage. So they began to search for alternate explanations; surely something was initially responsible for a scene that hadn't popped up anywhere else.

Some critics claimed that grunge emerged out of Seattle's physical geography, sandwiched as the city is between two mountain ranges, in a complex of lakes and islands themselves surrounded by the vast Puget Sound. Was the environment responsible for the noise and angst of grunge, or was it Seattle's famous Boeing Company, or both? Guns n' Roses bassist Duff McKagan once remarked in a 1994 interview with *Rolling Stone*: "You gotta understand Seattle, it's grunge. People are into rock & roll and into noise, and they're building airplanes all the time and there's lots of noise, and there's rain and musty garages."[6] Christopher Sandford, another *Rolling Stone* columnist, chalked up the sound to isolation: "[Seattle] has been called 'the hideout capital of the USA,' a far-flung outpost of a town . . . Seattle is physically remote."[7]

But if climate, topography, and industry were the catalysts for grunge, why didn't cohesive, similar sounds develop in other far Northwestern cities like Anchorage or Portland? Leaving environmentally determinist arguments aside, we discover that during the 1960s Seattle's University District and downtown had a thriving group of music venues. Bands that would frequent these venues were focusing their energies and attentions at the local level. After the height of punk culture in the mid-1970s and early 1980s, "underground" scenes and music cultures realized that local identity was their greatest asset. Throughout the '80s, this infrastructure attracted alternative bands to Seattle.

When the "grunge" sound solidified enough in the late 1980s to become a classifiable subgenre of alternative rock, it seemed that this theory worked and had produced something culturally significant.

As the grunge movement gained energy, it helped that there were a few record companies that could facilitate distribution and creativity without tarnishing the bands' underground image. Sub Pop, which had started as an underground music magazine in the early 1980s, and had turned into a Seattle record label a few years later, released the compilations *Sub Pop 100* and *Sub Pop 200* in 1986 and 1988, respectively, of songs from Seattle bands. Soundgarden signed with the label and released a single in 1987, followed by Nirvana and Mudhoney in 1988.

Though Sub Pop had tremendous luck in signing the right bands, it worked hard to maintain its underground image, and grunge bands worked hard to fit in with Sub Pop's desired "dark and muddy." By focusing on its local and underground identity, Sub Pop capitalized on the no-nonsense and anti-consumerist feel of the grunge bands.

Grunge's appearance on the national stage through MTV (especially Nirvana's *Smells Like Teen Spirit*) and in the popular news media ensured that it would no longer be confined to just one local scene. At the same time, the fashion world was turned upside-down by huge demand

7.8 "Thank you for the music," an informal memorial at a park bench near Kurt Cobain's Seattle home.

for items such as flannel shirts and torn jeans that grunge bands had originally worn. Like it or not, this fashion became one of the first hints of the commodified alternative-Seattle culture.

The straight-ahead, honest playing style and generally gloomy lyrics of grunge began to dominate alternative music. But as more bands across the nation began to adopt this style of playing, and the public began to demand music scenes that reflected the intimate, sweaty venues that incubated Seattle grunge, the narrative that Seattle bands had tried to keep local was exported and commercially exploited. Unfortunately, this was anathema to the central ethic of grunge. As Eric Iversen wrote of Nirvana in 1994, "The very 'Seattle' that they helped create has consumed Nirvana themselves, remade them into their own commodified image and effaced the original version under the veneer of domesticated deviance, sure to sell, but at the expense of their souls."[8] Grunge's fall was swift and brutal, and it dragged Seattle's musical prominence down with it until later in the decade when a more diversified, indie-oriented music infrastructure arose. The boom-and-bust cycle of music would only accelerate in the coming years.

In addition to the grunge movement, Seattle is also well known for its jazz culture. It has one of the few stations in the world to focus specifically on jazz (KPLU), and two of the city's high school jazz bands, Roosevelt and Garfield, frequently win or place highly in national championships. The city has produced some jazz greats, such as Quincy Jones, Ernestine Anderson, and Ray Charles. In addition to these three, other local artists were well known among the jazz greats of the time. These included the pianists Jimmy Rowles and Gerald Wiggins, who went on to play in Los Angeles; Patti Brown, who left for New York; bassists Wyatt Ruther and Buddy Catlett, who left to work with Count Basie; and artists such as Floyd Standifer who stayed on in the city. Famous jazz artists like Duke Ellington and Charlie Parker came frequently through Seattle and played at the "hub" of Jackson and Twelfth for decades, when it was the working-class section of the African American community.

Other than de Barros's masterful history of Seattle's jazz scene, there is little recorded history of the special jazz culture that developed in the city and that nurtured not only this art form but also the many music cultures that followed. In his words:

Jazz often thrived in Seattle, as elsewhere, in what were perceived as dives, in a black ghetto where gambling, prostitution, and illegal drinking were as central to the action as the music itself. The notion that something of cultural importance might be brewing outside the law, on the outskirts of respectability, was virtually inconceivable to the white reporters, editors, and cultural pundits who might have documented what was going on.[9]

The absence of this history in Seattle's cultural records is the result of racism and the fear of spaces of cultural difference; it is also the result of an East Coast bias, where those writing "culture" into the landscape often conspicuously avoided any mention of the strange and inconsequential places of the Pacific Northwest.

Diversity and Art in Seattle's Neighborhoods and Trails

FREMONT: AN ARTIST'S REPUBLIC

Once a small logging camp, Fremont is now one of the main bohemian and cultural hubs of Seattle. Accessible in the early days of the city by rail and trolley (running along what is now the Burke Gilman trail), Fremont was an ideal location for light industry. However, once the trolley and rail service ended in the late 1930s, Fremont's industrial prosperity declined, leaving the neighborhood a desolate, postindustrial space. As property values in the area fell, Fremont in the 1960s became a place for students, bohemians, and artists seeking inexpensive places to live. Gaining a reputation as an artist friendly space, Fremont's new culture began to form as more bohemians flocked to the neighborhood. Fremont has now officially declared itself an "Artists' Republic," and it holds the annual Summer

In recent years, technology companies such as Adobe and Getty Images have moved into the neighborhood, bringing renewed commercial investment. Inevitably, the new urban financing resulting from these major corporations has jeopardized the funkiness of Fremont with commodification and gentrification. Recent waves of gentrification have also caused the rapid transformation of other old neighborhoods, such as Belltown, Ballard, and Capitol Hill. Many of these areas are losing what made them unique, as the very spirit of "countercultural" or alternative lifestyles has proven attractive for marketing to a young, professional class.

CHINATOWN-INTERNATIONAL DISTRICT: AN ASIAN AMERICAN HUB

As one of Seattle's oldest neighborhoods, the International District, or ID, has a distinct and rich cultural history. It houses one of the most diverse Asian American settlements on the continent, in a space where Chinese, Filipinos, Japanese, Vietnamese, Koreans, and Cambodians have all settled together. The ID is located close to Pioneer Square, where mostly male Chinese settlers opened cigar shops, hotels, and restaurants, forming a hub for settlers who were hoping to work on the railroad lines or find a home during the Alaskan canneries' off-season. The first new Chinese residents were greatly resented by the larger white community, and immigrant communities were continually dispossessed of their land through racist policies. In February of 1886, when racial tensions ran especially high because of the economic recession, as many as three hundred Chinese settlers were forcibly evicted from their homes by an angry white mob. Even while the community eventually rebuilt, the original district was demolished for the expansion of Second Avenue. The current International District rests on tideflats, reclaimed and populated in the early 1900s.

Chinese immigration to Seattle began in the early 1860s, with mostly men congregating around Yesler's Mill at Elliott Bay. The community eventually moved to King Street by the early 1900s. In the late 1890s, direct steamship

7.9 Lenin in Fremont.

Solstice fair and parade. The self-proclaimed "Center of the Universe" (as it appears on a large Welcome sign), Fremont is considered to be one of the most eclectic and artistic neighborhoods in Seattle.

One of Fremont's best-known features is its collection of public art. Two popular sculptures, *Waiting for the Interurban* and a statue of Lenin, define Fremont's artist culture. The interurban sculpture, designed by Richard Byer, is a commemoration of the light rail Interurban line that once ran through the neighborhood. Locals often decorate the statue with ribbons, clothes, and costumes, a treasured community tradition. The Vladimir Lenin statue came to Fremont via Slovakia from artist Emil Venkov (fig. 7.9). Toppled in Slovakia's 1989 revolution, the statue was brought to Fremont by a Washington resident.

7.10 Seattle's Chinatown-International District.

routes between Yokohama and Seattle brought growing numbers of Japanese immigrants to Seattle. By the 1920s, the Japanese formed a Nihonmachi, or Japantown, largely along the steep slope of Main Street between Fourth and Seventh avenues. Filipinos entered the area after the Spanish American War under privileges to enter the United States as "nationals."This group of immigrants was dispersed in various hotels and flats in what became known as the International District. In the 1980s there was an influx of Southeast Asians seeking refuge after the Vietnam War. They formed a Little Saigon district, centered on the corner of Twelfth and Jackson. The 2000 Census put the Chinatown-International District population at 2,000 and marked it as one of Seattle's few ethnic neighborhoods.

Throughout the years, there have been episodes of racism, deportation, and confinement in the immigrant communities. As early as 1885, a white, working-class group, the Knights of Labor, agitated for the removal of all 350 Chinese from Seattle. On February 7, 1886, a group of whites corralled all the Chinese onto a dock at the foot of Main Street for deportation to San Francisco. The territorial governor blocked the forced removal, but 200 Chinese left voluntarily the following day. The remaining 150 Chinese

laborers refused to return to their homes, and a quickly mobilized white mob rioted. One person was killed and five were wounded.

The Nihonmachi had been thriving until the attack on Pearl Harbor in 1941. From 1942 to 1946, Japanese Americans were forcibly relocated to internment camps, causing many to lose their homes and businesses in Seattle (and elsewhere). Most of the Seattle internees were housed at Minidoka in Idaho. During World War II, many African Americans moved into the areas vacated by the Japanese. The post-war era also saw the displacement and dispossession of Asian Americans in the neighborhood. The construction of Interstate 5 cut a large swath through the heart of the district. In 1973 the construction of the Kingdome arena to the south also removed many businesses, hotels, and apartments. However, the neighborhood has maintained its integrity as an Asian American hub, not only through its vibrant markets and restaurants but also through its art, commerce, and constant promotion of local businesses.

On July 23, 1951, Seattle Mayor William F. Devin proclaimed the neighborhood the "International Center" and praised the mix of ethnicities. The claim of internationalism, however, was not welcome to everyone in the Chinese community; some considered it detrimental to their historical leadership there. In 1975, two groups clashed in separate competing applications for a public corporation, which would allocate City Council funds to develop a community center and house several social service agencies. One application came from the International District Improvement Association, (Inter*Im), a multiracial community group, and the other from the Chong Wa Benevolent Association, an organization that represented Chinese American residents. The Inter*Im application was to establish an "International District," while Chong Wa simply proposed a "Chinatown" Preservation and Development Authority. Mayor Ulhman created a compromise corporation, the "Seattle Chinatown/International District Public Development Authority," and to this day the name has stuck in the neighborhood, encompassing its true diversity, yet still recognizing the special history of the Chinese Americans.

7.11 A Washington State Ferry crosses Puget Sound.

FERRIES AND THE ISLAND LIFESTYLE

Seattle and the lifestyle and culture of its citizens have long been shaped by the mountains and waters of Puget Sound. Finding efficient ways of crossing the complex waters of the Sound has always been a priority. The area's first residents, including the Chinook, Makah, and Coast Salish peoples, were experienced and efficient sailors, crossing the waters in cedar log canoes. As later settlers began populating the region, negotiating the region's waterways became vital to the success of the new logging economy. Even as early as 1913, ferries began carrying trucks and cars across the Sound, connecting the young city of Seattle to the rich resources of the Olympic Peninsula.

What started out of economic necessity for early settlers has become a cultural symbol of Seattle and the Puget Sound. Although the public ferry system was intended to be a temporary substitute until a system of bridges could be built, the success and efficiency of the ferry system have turned it into a permanent and admired institution. Today, the Washington State Ferry system is the largest in the United States. In one year alone, the ferries carry on average over 10 million vehicles and 25 million passengers. By offering breathtaking views of the Seattle skyline and the Olympic Mountains, the ferries have become one of Seattle's most popular tourist experiences.

Beyond the popular appeal for tourists, the ferry system has left a lasting impression on the cultural landscape of Seattle. This distinctive system of travel has allowed for the development of unique subcultures within the Puget Sound region, namely, isolated island cultures that are created and sustained by the ferries. Even though they are surrounded by an urban metropolis of over 3 million people, islands such as Vashon retain a rural sensibility. Their secluded culture is treasured and passionately protected by Vashon's inhabitants. In late 1990, when the state government proposed a $2 billion bridge from the mainland to Vashon, more than a thousand protesters appeared at a local meeting to loudly condemn the bridge. For Vashon residents, a bridge to the small island would mean a sure end to their rural idyll, as commercial developments and expensive housing would no doubt take over. As one protestor wrote on an iconic sign, "If you build it, they will come." The ferry has made the island communities what they are today, small, quiet spaces where nature can be enjoyed, just out of range of a surging city.

The Olmsted Legacy

At the turn of the century, the City Beautiful movement in the United States did not bypass Seattle. Starting in 1903, the Olmsted brothers created a park system of rolling lawns and prominent boulevards out of what wilderness remained within the city. Their design aesthetic was based on the European tradition of "picturesque": pastoral landscapes and parks such as Volunteer Park and the Arboretum became highly groomed reflections of these idylls. Such an undertaking corresponded to the ruling mentality of the time, which associated Western Europe with the apex of civilization. The "beautification" program was not based primarily on the preservation of natural places in the city, so much as on the creation of an economic and social value and image.

While many residents of Seattle approved the construction of these parks, dissenters pointed out that the Parks Commission was placing the parks in upper-class areas in order to boost the neighborhoods' land values. The City Beautiful movement also worked in tandem with other major alterations to the landscape, including the regrading of Denny and Jackson streets. Along with the creation of the Olmsted Parks, these regrades permanently altered the topography of the city, generally to the advantage of a small group of investors and developers.

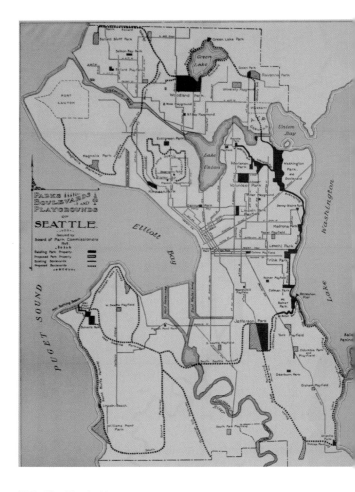

7.12   The Olmsted legacy.

THE BURKE-GILMAN TRAIL

The Burke-Gilman Trail was once a segment of the Seattle, Lake Shore & Eastern Railroad. The line was abandoned in 1970 and citizens rallied to turn it into a public trail for walkers, joggers, and bikers. By 1978 the first twelve miles of the Burke-Gilman multiuse trail had been built. Continued citizen action has helped extend the trail, segment by segment, to its present eighteen-mile length. Expansion of this trail continues in 2010 through community activism. A vigil was held in October 2009 to pressure officials to extend the trail along busy Shilshole Avenue; hundreds of people wore orange shirts and held signs declaring "We are the missing link."

The very concept of an urban trail system in Seattle suggests movement toward a renewed environmental consciousness in the city. The vast network of contemporary bicyclers using the trail seems to be a return to an earlier, more natural time, when commuters could use their own power to propel themselves through space. It is evident that the citizens of Seattle are aware of their power to reshape the spaces of the city, an awareness that became especially important in the world-changing struggles of 1999.

The Cultural Legacy of the WTO

"Whose streets? Our streets! Whose Streets? Our streets!" This chant captured the mood of the demonstrators who flooded the streets of downtown Seattle between November 30 and December 3, 1999. The World Trade Organization (WTO) was in town for its final ministerial meeting of the millennium, and over 50,000 protestors came out to voice their opinion: corporate-led globalization was offensive to them, and its defenders were not welcome in Seattle. Two opposing visions collided during that week, one driven by an ideology of the free reign of global capitalism, the other inspired by a belief that "another world is possible," promoted by a grassroots, democratic organization of the masses. Seattle was the perfect battleground for this fight: a trade-dependent city famous for its international footloose corporations, but also filled with liberal "greens" with a labor tradition that takes to the streets when push comes to shove.

Numerous struggles, both physical and symbolic, played out in the streets of Seattle: from battles over who controls something as large and complex as the global economy to battles over who controls something as small and seemingly mundane as the intersection of Sixth and Pine streets. From streets to the global economy, every country, every city, every house, every person, everything was a battleground.

Yet the tone of the protests was not simply a militant one. A more complex culture of resistance developed, one that very much included the concrete struggles for cultural control over the city. The WTO protests changed not just the physical landscape, but the mental and symbolic one as well. Many proponents of the anti-globalization movement now feel that the Seattle protest was the largest festival of resistance the world had yet seen.

Throughout the 1990s, Seattle saw an explosion of expensive developments within the downtown that were reputed to raise the "quality of life" in the city. Projects included the 1991 Seattle Art Museum ($61 million), the 1996 Nike Themetown Store ($25 million), the 1998 Benaroya Music Center and Concert Hall (118 Million), Pacific Place Retail-Cinema Complex and Parking Garage ($248 million), and the introduction in 1999 of not only Safeco Field ($517 million) but the promise of the upcoming Seahawks Stadium in 2002 ($430 million). The streets of the city were paved with a sense of symbolic modernization, cosmopolitanism, and progress. There was a certain smugness to the city, and a sense of inevitability about the future. The Battle of Seattle was shocking because it displayed discord and resistance in this vision of the future. It showed for the first time, a deep and powerful contestation over what and who could be represented in the political, economic, *and* cultural landscape of the city. Downtown had become a battleground.

This public resistance to "our" future city was contrary to the crafted image of Seattle in the 1990s. The city was a media darling, frequently topping numerous "best places to live" lists. Seattle's 1990s rise to prominence came during a time of great upheaval in the U.S. economy. As the United States faced the 1991 recession and an undercurrent of fear that it was losing its prominence in the global economy from a declining manufacturing base, Seattle was held up as a model city and a panacea for the ills facing the rest of the nation. With the expansion into the software side of the high-tech industry, Seattle was described as a "digital Paris" and appeared unscathed by the negative repercussions of the profound economic and social restructuring affecting other American cities.

It was no accident that Seattle came to host the WTO in 1999. As previous chapters have noted, Seattle always has been a city of global trade. Inviting the WTO to hold its ministerial meeting within metropolitan boundaries was yet another attempt by city officials and business elites to solidify Seattle's position as a global city. Seattle aspired to become the Geneva of the West—a small but world-class city hosting high-level meetings with significant outcomes.

The decision to host the WTO was part of a dominant discourse and vision about the city's future. But organizers failed to recognize the growing divisions within the city between those who had benefited from the immense wealth generated in Seattle in the past decades and those who were seeing their beloved city transformed into an unrecognizable and unlivable upscale and expensive playground for the few.

It was precisely this tension over who defines the economic and cultural geography of Seattle that was prominent in the WTO convention battle. When Seattle residents yelled "Whose Streets? Our Streets!" they were expressing a frustration rooted in watching decades of elite-led development of the city. From the World's Fair to the 1968 Forward Thrust and the widespread and growing corporate pervasiveness in neighborhoods, in the landscape, and in the skyline, the city seemed out of the control of "regular" folk. Thus, when Mayor Paul Schell said before the WTO gathering that the citizens of Seattle should be "tough on the issues, but gentle on *my* city" (our emphasis), Seattleites responded by letting it be known who should be represented in *everyone's* city.

NOTES

1   See judicial decisions *Tulee v. Washington* and *State v. Satiacum* for detailed information concerning these rulings.

2   Gabriel Chrisman provides a detailed history in "The Fish-in Protests at Frank's Landing," available through the Seattle Civil Rights and Labor History Project, http://depts.washington.edu/civilr/fish-ins.htm#_edn1.

3   Supreme Court Reporter, 1979, 3080. Quoted in Rita Brunn, "The Boldt Decision," *Law & Policy Quarterly* 4 (1982): 271–98

4   Richard Heyman, "Postindustrial Park or Bourgeois Playground," in *The Nature of Cities*, ed. M. Bennett and D. Teague (Tucson: University of Arizona Press, 1999).

5   James Lyons, *Selling Seattle* (London: Wallflower, 2004).

6   M. Azerrad, "Grunge City," *Rolling Stone*, April 16, 1992, 43–48.

7   Christopher Sandford, *Kurt Cobain* (New York: Carroll & Graf, 1996).

8   E. Iversen, "Brain Dead in Seattle," *The Baffler* 5 (1994): 21–25.

9   Paul De Barros and E. Calderon, *Jackson Street After Hours* (Seattle: Sasquatch Books, 1993).

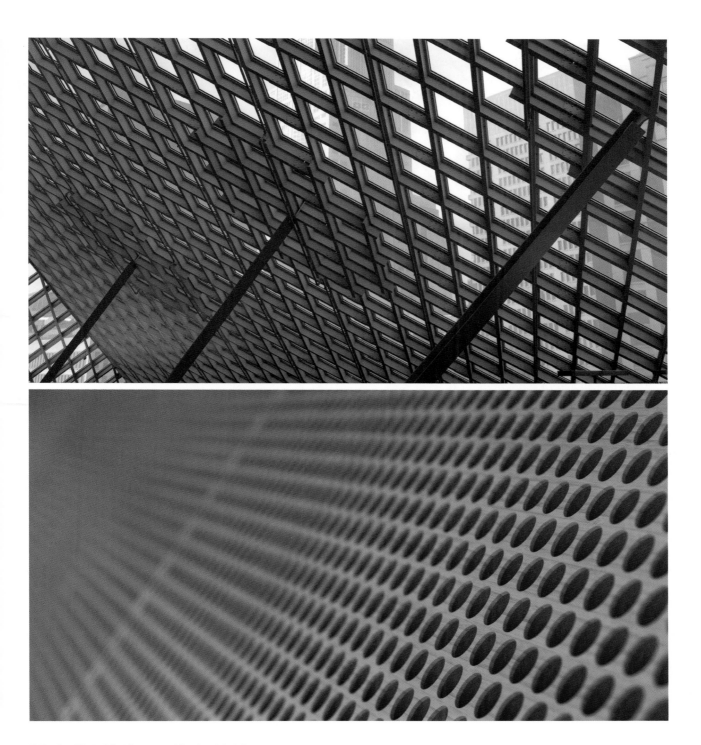

7.13 Seattle Public Library architectural details.

APPENDIX

APPENDIX 1. GEOGRAPHY'S SITUATION IN THE UW COLLEGE OF ARTS & SCIENCE

THE COLLEGE OF ARTS & SCIENCES

Arts	Humanities	Natural Sciences	Social Sciences
Art	Asian Languages & Literature	Applied Mathematics	American Ethnic Studies
Dance	Classics	Astronomy	American Indian Studies
Drama	Comparative History of Ideas	Biology	Anthropology
Digital Arts	Comparative Literature	Chemistry	Communication
Music	English	Mathematics	Economics
Burke Museum	French & Italian Studies	Physics	GEOGRAPHY
Henry Art Gallery	Germanics	Psychology	History
UW World Series	Linguistics	Speech and Hearing Science	Jackson School of International Studies
	Near Eastern Languages & Civilizations	Statistics	Law, Societies, and Justice
	Scandinavian Studies	The Burke Museum	Philosophy
	Slavic Languages and Literature		Political Science
	Spanish & Portuguese Studies		Sociology
	The Simpson Center for the Humanities		Women's Studies
			Burke Museum

APPENDIX 2. THE GEOGRAPHY DEPARTMENT FACULTY, 1927-2011

FACULTY, PhD AND YEAR GRANTED	APPOINTMENT
George T. Renner, Jr. Columbia, PhD, 1927	1927–33
Albert L Seeman, Washington, PhD, 1930	1928–42
Howard H. Martin, George Washington, PhD, 1929	1930–62
Frances M. Earle, George Washington, PhD, 1929	1932–67
Philip E. Church, Clark, PhD, 1937	1935–47
William E. Pierson, Washington, M.S., 1934	1937–46
John C. Sherman, Washington, PhD, 1947	1942–86
Mary Jo Read, Wisconsin, PhD, 1942	1943–45
Harold E. Tennant, Washington, PhD, 1937	1944–51
Dan Stanislawski, California, PhD, 1943	1945–47
Joseph E. Williams, Vienna, PhD, 1932	1946–50
Estelle A. Rankin, Columbia, M.A., 1935	1946–52
Marion E. Marts, Northwestern, PhD, 1949	1946–48, 1950–86
Graham H. Lawton, Oxford (U.K.), M.A., 1944	1947–51
William L. Garrison, Northwestern, PhD, 1950	1950–60
G. Donald Hudson, Chicago, PhD, 1934	1950–67
Henry M. Leppard, Chicago, PhD, 1928	1951–54
John D. Eyre, Michigan, PhD, 1951	1951–57
Edward L. Ullman, Chicago, PhD, 1942	1951–76
Rhoads Murphey, Harvard, PhD, 1950	1952–64
W. A. Douglas Jackson, Maryland, PhD, 1953	1955–94
Willis R. Heath, Washington, PhD, 1958	1957–72
George H. Kakiuchi, Michigan, PhD, 1957	1957–90
Morgan D. Thomas, Queen's (Belfast), PhD, 1951	1959–96
Richard L. Morrill, Washington, PhD, 1959	1960–97
Joseph Velikonja, Rome, PhD, 1948	1964–93
Richard A. Cooley, Michigan, PhD, 1962	1965–69
Ronald R. Boyce, Washington, PhD, 1961	1965–76
Douglas K. Fleming, Washington, PhD, 1965	1965–92
Phillip Bacon, Peabody, Ed.D., 1955	1966–71
Kuei–sheng Chang, Michigan, PhD, 1955	1966–87
Jacek I. Romanowski, Washington, PhD, 1970	1967–77
William B. Beyers, Washington, PhD, 1967	1967–2010
Phillip C. Muehrcke, Michigan, PhD, 1969	1969–73
Gunter Krumme, Washington, PhD, 1966	1970–2003
Jacob J. Eichenbaum, Michigan, PhD, 1972	1971–75

APPENDIX 2, CONTINUED.

FACULTY, PhD AND YEAR GRANTED	APPOINTMENT
Carl E. Youngmann, Kansas, PhD, 1971	1973–83
Virginia Sharp, Penn State, M.A., 1970	1974–79
David C. Hodge, Penn State, PhD, 1975	1975–2006
Craig Zumbrunnen, California (Berkeley), PhD, 1973	1977–
Jonathan D. Mayer, Michigan, PhD, 1977	1977–
Jaime C. Kooser, California (Berkeley), PhD, 1981	1980–85
Timothy L. Nyerges, Ohio State, PhD, 1980	1985–
Victoria A. Lawson, Ohio State, PhD, 1986	1986–
Nicholas R. Chrisman, Bristol (UK), PhD, 1982	1987–2004
Lucy Jarosz, California (Berkeley), PhD, 1990	1990–
Kam Wing Chan, Toronto, PhD, 1988	1991–
Katharyne Mitchell, California (Berkeley), PhD, 1993	1993–
Matthew Sparke, UBC, 1996	1995–
J. W. Harrington, Washington, PhD, 1983	1997–
Nayna Jhaveri, Clark University, ABD, 1996	1997–2003
Suzanne Davies Withers, UCLA, PhD, 1992	1997–
Michael Brown, UBC, PhD, 1994	1998–
Mark Ellis, Indiana, PhD, 1988	1999–
Kim V. L. England, Ohio State, PhD, 1988	1999–
Steven K. Herbert, UCLA, PhD, 1995	2000–
Craig Jeffrey, Cambridge (UK), PhD, 2006	2005–9
Sarah Elwood, Minnesota, PhD, 2000	2006–

APPENDIX 3. TENT CITY LOCATIONS, 1998-PRESENT

TENT CITY 3 HOSTS

King Vn, Inc
3011 S. Estelle
Seattle, WA 98144

Colman School

Shower Of Blessings Tabernacle
C/O 3818 Edmunds S. #307
Seattle, WA 98108

All Saints Episcopal Church
5150 S. Cloverdale
Seattle, WA 98118

Rainier Avenue Free Methodist
5900 Rainier Ave S
Seattle, WA 98118

Crown Hill United Methodist
8500—14th Ave NW
Seattle, WA 98117

Seattle Advent Christian Church
(Now Defunct)
1300 E Olive
Seattle, WA 98122

St. Marks Episcopal Cathedral
1245 10th Ave E
Seattle, WA 98102

Star Bethel Missionary Baptist Church
5922 Rainier Ave S.
Seattle, WA 98118

Northgate Baptist Church
10510 Stone N
Seattle, WA 98133

El Centro De La Raza
2524 16th Ave S
Seattle, WA 98144

Martin Luther King, Jr Park

Riverton Park United Methodist
3118 S 140th
Tukwila, WA 98168

Trinity United Methodist
6512 23Rd NW
Seattle, WA 98117

Lake City Christian Church
1933 Ne 125th
Seattle, WA 98125

North Seattle Church Of The Nazarene
13130 5th Ave Ne
Seattle, WA 98125

St. Therese Catholic Church
3416 E. Marion
Seattle, WA 98122

Dunlap Baptist Church
(Sold Their Land, Now Defunct)
8445 Rainier Ave S
Seattle, WA 98118

Church By The Side Of The Road
S. 148th & Pac Hwy S
Tukwila, WA 98188

Haller Lake United Methodist Church
13055 First Ave Ne
Seattle, WA 98125

Highline United Methodist Church
13015 First Ave S
Burien, WA 98168

Phinney Ridge Lutheran Church
7500 Greenwood Ave N
Seattle, WA 98103

Shoreline Free Methodist
510 Ne 170th
Shoreline, WA 98155

Temple Beth Am
2632 NE 80th
Seattle, WA 98115

Cherry Hill Baptist Church
22Nd & Cherry
Seattle, WA 98122

St. Joseph's Catholic Church
732—18th Ave E
Seattle, WA 98112

Seattle University
14th & Cherry

St. George Episcopal Church

Vacant Lot
Martin L King Jr WAy S & S 129th St
Seattle WA 98178

CONTRIBUTORS

VOLUME EDITORS

MICHAEL BROWN is Professor of Geography at the University of Washington. He is an urban political and cultural geographer who studies sexuality and health in Seattle.

RICHARD MORRILL is Emeritus Professor of Geography at the University of Washington. He is a human geographer (still hard at work), with particular interests in population, inequality, and urban and regional planning.

CONTRIBUTORS

DAVID BARKER received his B.A. in Geography in 2007 from Middlebury College, where he researched demographic and economic trends in the Interior Columbia River Basin with Peter Nelson. He is the Deputy Director of PlaNYC at the New York City Department of Parks and Recreation.

MIKAIL AYDYN BLYTH is an honors undergraduate student in geography at the University of Washington.

ANNE BONDS is Assistant Professor of Geography at the University of Wisconsin–Milwaukee. She is an economic geographer with research interests in political economy, social theory, and geographies of poverty and inequality. She received her PhD in Geography from the University of Washington in 2008.

ETHAN BOYLES received a double B.A. degree with honors in Geography and Economics from the University of Washington in 2010. He is interested in political geography, with a focus on construction of and representation in urban landscapes.

WILLIAM B. BEYERS is Professor of Geography at the University of Washington. His research has focused on structural change in regional economies and on the development of the service economy.

JOHN CARR is Assistant Professor of Geography at the University of New Mexico. He is an urban political and legal geographer who studies the politics of public space. He graduated with a PhD from the University of Washington in 2007.

KAM WING CHAN is Professor of Geography at the University of Washington. His research interests include China's urbanization, economy, migration, and cities.

SPENCER COHEN is a PhD candidate in Geography at the University of Washington. His interests are in local government finance, political economy, and land use in modern China. He also currently works as the research manager for the Washington State Department of Commerce.

JENNIFER DEVINE is a PhD candidate at the University of California, Berkeley, and a University of Washington undergraduate alumna in Geography. She studies the socio-spatial politics of tourism development and environmental conservation in Guatemala.

MARK ELLIS is Professor of Geography at the University of Washington. He is a population geographer who studies immigration, ethnicity, racialized identities, and labor markets in the United States.

SARAH ELWOOD is Associate Professor of Geography at the University of Washington. Her work integrates GIScience, urban political geography, and mixed methods.

KIM ENGLAND is Professor of Geography at the University of Washington. She is an urban social and feminist geographer whose research focuses on care work, critical social policy analysis, economic restructuring, and inequalities in North America.

SOFIA GOGIC received her B.A. with honors in Geography from the University of Washington in 2010. She is interested in applying what she has learned to the field of urban design and planning.

JAMES W. ("JW") HARRINGTON is Professor of Geography at the University of Washington. He specializes in subnational, regional economic development, through studying labor and workforce development, producer services, and international trade.

STEVE HERBERT is Professor of Geography and Law, Societies, and Justice at the University of Washington. Much of his work focuses on the politics of public safety and the practices of policing in Seattle.

LUCY JAROSZ is Associate Professor of Geography and Adjunct Associate Professor of Women Studies at the University of Washington. She studies and teaches about hunger, poverty, food, and the political economy of agricultural development In Seattle and beyond.

CHARLES W. KAUFFMAN is an undergraduate student in Geography at the University of Washington. As an economic geographer, he studies regional economic development, international trade, and retail.

LARRY KNOPP is Professor and Director of Interdisciplinary Arts & Sciences at the University of Washington–Tacoma, and Adjunct Professor of Geography at the University of Washington (Seattle). He is a social, political, and cultural geographer, with research interests in sexuality, cultural politics, and elections.

JOHN ERIC KRAMAK is a policy development intern at Wallingford Solar Initiative. He received his B.A. in Geography with honors from the University of Washington in 2010, and his studies focus on the conflictual geographies of culture, the environment, and housing in the Seattle area. He is looking forward to graduate studies in urban geography/planning at the University of Cape Town, South Africa.

VICTORIA LAWSON is Professor of Geography at the University of Washington. She is a feminist development geographer who studies poverty across the Americas.

RITA B. LEE received her B.A. with honors in Geography and Economics from the University of Washington in 2010. Her studies focus on utilizing GIS to analyze urban planning and development. She plans to hold off graduate school until she is tired of traveling.

KATHARYNE MITCHELL is Professor and Chair of the Department of Geography at the University of Washington. She is interested in questions of education, citizenship, and culture and has spent many years studying these themes in the context of contemporary migration and rapid urban change.

LISE NELSON is Associate Professor of Geography at the University of Oregon. She is a political and cultural geographer who studies spatialities of gender, race, and belonging in Mexico and the rural U.S. in the context of neoliberal globalization. She received her PhD from the University of Washington in 2000.

PETER NELSON is Associate Professor of Geography at Middlebury College. He is an economic geographer studying the intersection of rural economic and demographic change in the United States. He received his PhD from the University of Washington in 1999.

TIMOTHY NYERGES is Professor of Geography at the University of Washington. His research is about participatory geographic information science and systems enabled by cyberinfrastructure technology to support stakeholder-based problem solving for land use, transportation, and water resources on the global coast.

HAYLEY PICKUS received her B.A. with honors in Geography, with a minor in Spanish, from the University of Washington in 2010. She is interested in global health and plans to pursue a Master of Public Health degree.

KEVIN RAMSEY received his PhD in Geography from the University of Washington in 2009. His research and teaching intersects urban political geography, environmental studies, and "critical" geographic information science.

GEORGE ROTH is an undergraduate student at the University of Washington, who is scheduled to receive a B.S. with honors in Oceanography, with a minor in Geography, in 2012. He studies geographic information systems, remote sensing, climate change, and the Arctic.

TRICIA RUIZ is a PhD candidate in Geography at the University of Washington. Using critical race theory and spatial demography, she studies school segregation and educational reform in the United States.

GARY SIMONSON is an urban and cultural geographer who studies gentrification and development patterns in Seattle. He recently received his M.A. in Geography from the University of Washington.

MATTHEW SPARKE is Professor of Geography and International Studies and Adjunct Professor of Global Health at the University of Washington. His most recent work is focused on the ties and tensions between global health and globalization.

TONY SPARKS received his PhD in Geography from the University of Washington in 2009 and currently works as a lecturer at Sonoma State University. He studies the relationships between social policy, poverty, homelessness, and urban informality.

ANNE STEINBERG received her B.A. with honors in Geography from the University of Washington in 2010. Her interests include international development studies and critical development theory. She intends to pursue a degree in law.

NICOLE S. STRAUB is an honors undergraduate student in Geography at the University of Washington.

LOLA S. STRONACH is a graduate student at the University of Washington, working toward her M.P.H. in Nutritional Sciences, along with certification as a registered dietitian. Her studies focus on GIS, food, health and inequality. She received her B.A. with honors in Geography and her B.S. in Public Health from the University of Washington in 2010.

CARL URNESS received a double B.A. degree with honors in Asian Studies and Geography from the University of Washington in 2010. He is attending graduate school for a Master of Public Affairs degree.

NICHOLAS VELLUZZI received his PhD in Geography from the University of Washington in 2007. He is an economic geographer, with research interests in regional production systems, globalization, and political economy.

CATHERINE VENINGA holds a PhD in Geography from the University of Washington. She is an urban geographer whose research focuses on politics and race in the city. She lives in Brooklyn, New York, with her husband and daughter.

SEAN WANG is an undergraduate studying social and cultural geographies, as well as digital humanities and public scholarship, at the University of Washington. He is spending 2010 on an exchange fellowship at the University of Edinburgh, Scotland, and will receive his B.A. in Geography in 2011.

MATTHEW W. WILSON is Assistant Professor of Geography at Ball State University. His work in technology studies draws upon urban political geography and critical GIScience. He holds a PhD in Geography from the University of Washington.

SUZANNE DAVIES WITHERS is Associate Professor of Geography at the University of Washington. She is a population and urban geographer who studies family migration, intergenerational mobility, and housing markets from a longitudinal perspective.

CREDITS

*Thanks to the following individuals and agencies for their
contributions of photographs, maps, charts, and diagrams.*

PHOTOGRAPHS

Sophia Agtarap: p. xvi

Anne Bonds: fig. 4.5

Michael Brown: figs. 2.9, 2.17, 5.15b, 6.44, 6.45, 6.49, 6.51,
6.52, 7.1, 7.2 7.8, and 7.10

Jeff Dietz: fig. 1.8

Sarah Elwood: fig. 5.19

Virgil Gloria: pp. v, xi

James Goldsmith: pp. viii, xiii, and 184; fig. 1.4

Andrew Gorohoff: pp. x, xii, xiv, xv, xvii, xviii, 195, and 196;
figs. 1.1, 1.2, 1.3, 1.5, 2.17, 2.18, 7.11, and 7.13

Dan Hughes: figs. 5.8 and 5.9

Andrew Larsen, under Creative Commons license: fig. 6.41

Vicky Lawson: figs. 4.7, 4.9, and 4.10

Joe Mabel: figs. 7.3 and 7.7

Katharyne Mitchell: fig. 7.4

Jehane Ramola: fig. 4.4

Kevin Ramsey: fig. 5.15a

Matthew Sparke: figs. 3.1–3.5

Tony Sparks: figs. 6.22, 6.25, and 6.26

Museum of History and Industry: figs. 6.27 (PI25526) and
6.28 (P125782)

Northwest Lesbian & Gay History Museum Project: fig. 6.54

Oregon Historical Society, Valley Migrant League Photograph Collection: fig. 4.8

Seattle Municipal Archives: figs. 6.42 (document collection,
1171) and 6.43 (record series, 1628–02, 32234), and 7.12
(Map 607)

University of Washington Libraries, Special Collections:
figs. 2.13 (oversize G3521.P1 1905.W5), 2.14 (HMJ0277),
6.46 (UW5878), 6.47 (LEE058), and 7.6 (A. Curtis 23588)

MAPS

Mike Babb: figs. 1.5, 1.6, 5.10, 5.20, and 6.40

Mike Babb, based on maps originally created by
Richard Morrill: figs. 6.9–6.13

Mike Babb, based on maps originally created by the Seattle
Civil Rights and Labor History Project: figs. 6.32 and 6.33

Michael Brown: fig. 6.50

Michael Brown and Mike Babb: figs. 6.52, 6.55–6.57

Richard Morrill and Mike Babb: figs. 2.8, 2.10–2.12, 5.1,
5.3–5.6, 6.2, 6.4–6.7

Richard Morrill, Mike Babb, and Lola Migas: figs. 2.7, 6.1

Lise Nelson: fig. 4.6

Pete Nelson: figs. 4.1–4.3

Tricia Ruiz: figs. 6.29, 6.37–6.39

Gary Simonson and Mike Babb: fig. 6.48

Catherine Veninga: fig. 2.15

Sean Wang and Mike Babb: fig 1.9

Matthew Wilson: figs. 5.7 and 5.16

Suzanne Davies Withers and Mike Babb: figs. 6.16–6.21

King County: fig. 5.2

Puget Sound Regional Council: fig. 6.3

Washington Global Health Alliance: fig. 3.6

CHARTS AND DIAGRAMS

Derek Andreoli and William Beyers: fig. 2.5

William Beyers and Lola Migas: figs. 2.1–2.3

Lola Migas, based on charts originally created by the
    Port of Seattle: figs. 2.6 and 2.19

Tricia Ruiz: figs. 6.34 and 6.35

Tricia Ruiz and Mike Babb: figs. 6.30 and 6.31

Suzanne Davies Withers: figs. 6.14 and 6.15

The Seattle Times: fig. 2.4

INDEX